The BATTLE *for*
CHU MOOR
MOUNTAIN
VIETNAM, APRIL 1968

The BATTLE *for* CHU MOOR MOUNTAIN
VIETNAM, APRIL 1968

As told by the soldiers who were there

FRED CHILDS

DEEDS PUBLISHING | ATLANTA

Published by Deeds Publishing, Marietta, GA
www.deedspublishing.com

Library of Congress Cataloging-in-Publications Data is available upon request.

ISBN 978-1-941165-43-0

Books are available in quantity for promotional or premium use. For information, write Deeds Publishing, PO Box 682212, Marietta, GA 30068 or info@deedspublishing.com.

10 9 8 7 6 5 4 3 2 1

DEDICATED TO

The soldiers who fought and sacrificed on
the Vietnam battlefield of Chu Moor Mountain
in April 1968

&

The Army's 4th Infantry Division past, present, and future

TABLE OF CONTENTS

PRELUDE

AT 1602 HOURS, ON 24 APRIL 1968, THE RADIO CRACKLED OUT A message to the nearby fire support base, "D Company requests a basic load of ammo." Eleven short minutes later, a more urgent message was sent, "D Company needs ammo ASAP!" In this short period of time, it had become quite evident that a large enemy force had been contacted and a battle of serious proportions was heating up. By late in the evening, ten U.S. Army soldiers had been reported wounded, a re-supply helicopter had been shot down, and D Company appeared to be surrounded by elements of the North Vietnamese Army's (NVA) 66th Regiment. Within a week, scores more would be wounded and killed on both sides. D Company was part of the 1st Battalion, 22nd Infantry Regiment (D/1/22), 4th Infantry Division, and was the Battalion's lead unit as it first reached the summit of Chu Moor Mountain, located in the western Central Highlands in Vietnam.

CHAPTER 1
Steadfast and Loyal

THE UNITED STATES ARMY'S 4TH INFANTRY DIVISION DATES BACK to the First World War where it was first organized at Camp Greene, North Carolina on 10 December 1917. The Division quickly adopted their distinctive unit insignia, the four ivy leaves. This came from the Roman numerals for four (IV) and signified their motto of "Steadfast and Loyal." Ivy symbolizes tenacity and fidelity. Their second nickname, "Iron Horse," has recently been adopted to indicate the speed and power of the Division now that it is mechnized.

In WW I, the Division participated in both the St. Mihiel Offensive and the two phases of the Meuse-Argonne Campaigns. During the war, it suffered 2,611 killed in action and 9,895 wounded. After occupational duty in Germany, it was inactivated in 1921 at Camp Lewis, Washington. Reactivated again in 1940 at Fort Benning, Georgia prior to WWII, the Division took part in the Normandy invasion and landed at Utah Beach on 6 June 1944. It would later fight at Sainte-Mere-Eglise, Cherbourg, Paris, and the Siegfried Line at Schnee Eifel, the Battle of the Huertgen Forest in Germany, and the Battle of the Bulge in Luxemburg. It eventually crossed the Rhine into Germany and fought until the end of the war, when it was sent home to prepare to fight in Japan. The 4th's casualties for the war included 4,097 killed, 17,371 wounded, and 757 who later died of wounds. The division was inactivated in early 1946 while serving at Camp Butner, NC.

Reactivated in 1947, the Division was sent to Germany in 1950 as the first division to return for duty in the Cold War, standing strong against the Russian threat. After a five-year tour in Germany, where it was head-quartered in Frankfurt, it returned to Fort Lewis in 1956.

Led by the 2nd Brigade in July, 1966, the 4th Infantry Division was deployed to Vietnam where they began a deployment of over four years on 25 September 1966. Throughout its service in Vietnam, the Division conducted combat operations in the western Central Highlands along the borders between Vietnam, Laos, and Cambodia. It fought many battles against the NVA in the mountains surrounding Kontum and Pleiku, and was later involved in the cross-border operations during the Cambodian Incursion (April-July 1970).

The Division set up their Vietnam headquarters, south of Pleiku City at Camp Enari, named after the 4th Division's first posthumous Silver Star recipient, Lieutenant Mark Enari. Camp Enari, the 4th Infantry Division's main base camp while in Vietnam in the later 1960s, stretched from Dragon Mountain to the northwest to the Hensel Army Airfield in the south. During their last year in Vietnam, in 1970, the Division worked out of the former 1st Cavalry Division base camp at An Khe.

THE DIVISION'S UNITS DURING THE VIETNAM WAR INCLUDED THE FOLLOWING

1st Battalion, 8th Infantry
2d Battalion, 8th Infantry (Mechanized)
3d Battalion, 8th Infantry
1st Battalion, 12th Infantry
2d Battalion, 12th Infantry (to 25th ID, Aug 67-Dec 70)
3d Battalion, 12th Infantry
1st Battalion, 14th Infantry (from 25th ID, Aug 67-Dec 70)
1st Battalion, 22d Infantry (Separate, Nov 1970 to Jan 1972)
2d Battalion, 22nd Infantry (to 25th ID, Aug 67-Dec 70)
3d Battalion, 22nd Infantry (to 25th ID, Aug 67-Dec 70)
1st Battalion, 35th Infantry (from 25th ID, Aug 67-Apr 70)
2d Battalion, 35th Infantry (from 25th ID, Aug 67 to Dec 70)

2d Battalion, 34th Armor (to 25th ID, Aug 67-Dec 70)

1st Battalion, 69th Armor (from 25th ID, Aug 67 to Apr 70)

2nd Battalion, 9th Artillery (105 mm) (from 25th ID, Aug 67 to Apr 70)

5th Battalion, 16th Artillery (155 mm)

6th Battalion, 29th Artillery (105 mm)

4th Battalion, 42d Artillery (105 mm)

2d Battalion, 77th Artillery (105 mm) (to 25th ID, Aug 67 to Dec 70)

1st Squadron, 10th Cavalry (Armored) Division Reconnaissance

4th Aviation Battalion

4th Engineer Battalion

4th Medical Battalion

124th Signal Battalion

704th Maintenance Battalion

Company E, 20th Infantry (Long Range Patrol)

Company E, 58th Infantry (Long Range Patrol)

Company K, 75th Infantry (Ranger)

4th Administration Company

4th Military Police Company

374th Army Security Agency Company

The cost to the Division during the Vietnam War was 2,511 dead and 15,229 wounded.

Twelve soldiers earned the Medal of Honor, ten of whom were awarded posthumously.

Name	Rank	Unit	Date
BELLRICHARD, Leslie Allen **	PFC	C/1/8	5/20/67
BENNETT, Thomas W. **	CPL	B/1/14	2/11/69
EVANS, Donald W. Jr. **	SP/4	A/2/12	1/27/67

GRANDSTAFF, Bruce Alan **	P/SGT	B/1/8	5/18/67
JOHNSON, Dwight H	SP/5	B/1/69	1/15/68
McDONALD, Phill G. **	PFC	A/1/14	6/7/68
MCKIBBEN, Ray **	SGT	B/7/17th CAV	12/8/68
McNERNEY, David H.	1/SGT	A/1/8	3/22/67
MOLNAR, Frankie Z. **	S/SGT	B/1/8	5/20/67
ROARK, Anund C. **	SGT	C/1/12	5/16/68
SMITH, Elmelindo R. **	S/SGT	C/2/8	2/16/67
WILLETT, Louis E. **	PFC	C/1/12	2/15/67

** Posthumous Awards

Today, the Division continues to battle, fighting in the 'War on Terrorism,' having served in Iraq during Operation Iraqi Freedom (OIF) and Operation New Dawn, and in Afghanistan during Operation Enduring Freedom (OEF). On 13 December 2003, the 4th Infantry Division earned the distinction of being credited with the capture of Iraq's President, Saddam Hussein. Two 4th Division soldiers earned the Medal of Honor in Afghanistan.

CHAPTER 2
Location, Location, Location

DURING THE AMERICAN PRESENCE IN SOUTH VIETNAM IN THE 1960's and 1970's, the country was divided into four tactical zones from north to south: I Corps or I CTZ (Corps Tactical Zone), II Corps, III Corps, and the IV Corps. The II CTZ was the largest in size, comprising some 84,000 square kilometers or roughly 45% of the land mass of South Vietnam. It is roughly the size and shape of the state of Florida. It is bounded on the west by Cambodia and Laos, the South China Sea to the east, the I CTZ (Quang Tin and Quang Nhai Provinces) to the north and the III CTZ (Phuoc Long, Long Khanh and Binh Tuy Provinces) to the south.

There are three main types of terrain within the II Corps area: the lowlands, the plateaus and the mountains. The plateau region was known as the Central Highlands (Tay Nguyen) and represented approximately 51,800 square kilometers of rugged mountain peaks, extensive forests and rich soil. It was located along the western portion of the II CTZ from the Darlac Province on the south, though the Pleiku Province and ending in the Kontum Province to the southern border of the I CTZ in the north.

The city of Kontum was situated in the middle southern portion of Kontum Province. Several miles to the west of the city, lay the northeastern border tip of Cambodia as it met Laos and South Vietnam in a narrow finger-like projection. Between Kontum City and just a few miles from Cambodia, Chu Moor Mountain lay in wait for whoever wanted to control the surrounding area.

The population in the II CTZ was approximately 2.8 million, including over one-half million indigenous Montagnards, most of whom lived in the Central Highland area provinces. There were very few roads in the Central Highlands, with only the major cities connected north to south by Route 14 and east to west by Route 19. The weather in the II Corps consisted of two seasons—wet and dry. During the summer monsoon, which runs from June to September, the winds are from the southwest and the mountains and Central Highlands have low ceilings with lots of rain, while the coastal areas are dry. The winter monsoon, from September to April, bring winds from the northeast causing coastal rains and easing, but not totally eliminating, the harsh rain conditions in the plateau regions.

The North Vietnamese recognized early on that the Central Highlands were a strategic area of major importance, essential to control and to dominate South Vietnam. Its close and adjacent access to sanctuaries in Cambodia and Laos was an added benefit for the North Vietnamese Army (NVA). In 1965, the NVA attempted to cut Vietnam in two with an attack from its Cambodian sanctuaries across the Central Highlands and on to the sea. Their plan was thwarted by the U.S. First Cavalry Division in the famous battle of the Ia Drang Valley (the first major fight between the Americans and NVA) in November 1965. Ten years later, in 1975, the NVA used this same strategy and within just a few weeks, the country was cut in half and Saigon quickly fell and ended the war.

It was in these Central Highlands of the Pleiku and Kontum Provinces that the 4th Infantry Division took over the main responsibility from 1966-1970.

CHAPTER 3
Operation MacArthur

DURING THE SUMMER OF 1967, HEAVY CONTACT WITH NVA FORCES in the Central Highlands of the Kontum Province prompted the Americans to launch Operation Greeley on 16 June 1967. This was a combined search and destroy effort by elements of the 4th Infantry Division, 173rd Airborne Brigade, and the Army of the Republic of Vietnam (ARVN)'s 42nd Infantry Regiment and Airborne units. The fighting remained intense until sometime in August when it appeared that the NVA had withdrawn its forces. Greeley was officially ended on 12 October 1967.

By October 1967, American intelligence indicated that the communist forces had been reinforced and combined into the 1st People's Army of Vietnam (PAVN) which was another name for the NVA. It appeared that this force was directed to take the city of Dak To and the surrounding area. To combat this new threat, Operation MacArthur began immediately on 12 October 1967.

Operation MacArthur was a long term operation that was overseen by the 4th Infantry Division. It lasted from 12 October 1967 to 31 January 1969. It was a basic search and destroy operation consisting of border protection and reconnaissance in force searches designed to eliminate the communist's infiltration from Cambodia and Laos and to find those already within the region. By the time the operation ended, some 700 plus U.S. soldiers would be killed and over 1,400 wounded. The NVA would have 5,731 battle casualty deaths.

The first and probably most famous battle of the operation became known as the Battle of Dak To which took place from 3-22 November 1967.

One of the bloodiest battles of the war took place in the Central Highlands near Dak To. About 4,500 troops of the US 4th Infantry Division and 173rd Airborne Brigade faced off against 6,000 North Vietnamese troops of the 174th regiment. The North Vietnamese were forced to withdraw with 1,455 dead troops. US casualties numbered 285 killed and 985 wounded.

The next significant battle during Operation MacArthur was the First Battle of Kontum during the now infamous TET offensive of 1968 (the Second Battle of Kontum occurred during the 1972 Easter Offensive). The battle began on 30 January and lasted through 12 February 1968. Initially, the NVA and Viet Cong (VC) attacked the Provincial Capital of Kontum City but they were prevented from taking total control of the city by elements of the South Vietnamese Army. The enemy did, however, take and maintain control of several large areas of the city and the surrounding countryside. 4th Infantry Division units under its new Commander, Major General Charles Stone, quickly came to the rescue and were given the task of clearing the enemy out of the city and pursuing him in the outlying territory.

The commanding officer of the 1st Battalion, 22nd Infantry was placed in charge of the battle and was directed to assume command of all 4th Infantry Division forces in the area during the 14-day fight.

After the Battle of Kontum, Operation MacArthur continued in the Central Highlands with numerous battles, both large and small. One such battle was fought for Chu Moor Mountain, located just a few miles from the Cambodian border. It was situated in a valley that had seldom, if ever, seen American troops in force. Once again, the 1st Battalion, 22nd Infantry would be the lead element in the fighting, although they would be supported by numerous other units, including:

GROUND UNITS

1st Battalion 22nd Infantry	1st Battalion 14th Infantry	1st Battalion
Company A	Company A	12th Infantry
Company B	Company B	Company A
Company C	Company C	
Company D	Company D	
Company E		
Headquarters Company		

1st Battalion 8th Infantry
Company B

AVIATION UNITS

57th AHC (Cougar Guns & Gladiator Slicks) A Troop 7/17 Cavalry	119th AHC (Gator Slicks) Co B 4th AVN BN (Gambler Guns)	A Troop 1/9 Cavalry (Headhunters)
A Co 158th AVN BN (Ghost riders)	Dustoff 33, 34, 36, 37 and 38 (Dustoff was call sign for medical evacuation choppers)	

ARTILLERY SUPPORT

Engineers

C Battery 5/16 Artillery
C Battery 5/22 Artillery
C Battery 4/42 Artillery
B Battery 1/92 Artillery
C Battery 1/92 Artillery

B Company 4th Combat Engineers

UNITED STATES AIR FORCE

Cider FACS (Forward Air Control)	Spooky 21, 22 and 23 (AC-47 gunships)	Arc light (B-52 bomber strikes)

The colorful history of the 22nd Infantry Regiment dates back all the way to the War of 1812 where it was part of General Winfield Scott's Brigade. Due to a lack of proper uniforms, the Regiment was mistaken as local militia during the battle against the British at Chippewa, Canada in 1814. The British commander, General Riall, at first dismissed their force as a serious threat, but when he saw them attacking in good order through artillery and musket fire, he realized his error and shouted, "Those are regulars, by God." Henceforth, the unofficial motto of the Regiment became "Regulars by God." Their official motto is "Deeds Not Words."

After the War of 1812, the Regiment fought in the Indian Wars from 1866-1898. The next combat came in the Spanish American War of 1898 and the Philippine Insurrection in 1899-1905. Although it did not fight in the First World War, it had earlier provided assistance during the 1906 San Francisco earthquake. The Regiment fought with distinction during the Second World War, initially landing in France on D-Day, and fighting all the way to Germany. During the Cold War from 1951-1956 it was stationed back in Germany, first being head-quartered at Schweinfurt, then at Giessen, and finally at Kirchgoens.

The next duty station for the 22nd Infantry was at Fort Lewis, Washington from 1956-1966.

It was then deployed to Vietnam and served for six years until returning to the United States in early 1972 where it was assigned to Fort Carson, Colorado. During its Vietnam service, the soldiers of the Regiment's 1st Battalion (1/22) fought in 13 campaigns. The Battalion stayed with the 4th Infantry Division until the division left Vietnam, then 1/22 came under the command of the First Field Force Vietnam (IFFV) until returning to the States in late January 1972.

It was the longest Vietnam serving unit within the 4th Infantry Division. During the Tet Offensive, 1/22 earned the Valorous Unit Award for its fighting accomplishments in Kontum City.

It was also awarded two, Republic of Vietnam Cross of Gallantry Unit Citations.

CHAPTER 4
Prelude to the Battle of Chu Moor Mountain

"The enemy lived underground like moles and
remained hidden most of the time."

BY APRIL 1968, THE AREA AROUND CHU MOOR MOUNTAIN HAD seldom, if ever, seen American troops in force. There had been several recent reports that aircraft flying near the mountain had taken enemy anti-aircraft fire. It appeared that this was an area worth investigating and to perhaps engage the enemy on the Vietnam side of the border, as the offensive, countryside strategy of the MACV Commander, General William C. Westmoreland, dictated. Therefore, the 4th Infantry Division's Commanding Officer, Major General Charles P. Stone, ordered some of his Division's assets into the area.

Chu Moor Mountain lay in a valley in the II Corp Tactical Zone's Central Highlands. It was west of Kontum City and just a few miles east of the Cambodian border where it formed a narrow finger, meeting both Vietnam and Laos. The Ho Chi Minh Trail crossed into Vietnam in the immediate vicinity.

The surrounding valley and mountain were heavily vegetated with jungle and large trees. Specialist-4 and senior medic, Bud Roach, with C/1/22 recalled, "The mountain was really a peak in a ridge line running north and south. There was a saddle on the south side of the peak that rose to another crest not quite as high. The hillside was very steep and overgrown with dense, double or triple canopy forest. The

NVA must have been using the area as a staging point off the Ho Chi Minh trail for many years. The entire vicinity was crawling with well-equipped and dug in NVA regulars. They had tunnels, spider holes, and snipers tied in the trees."

Corporal Doug Stanek described, "The enemy lived underground like moles and remained hidden most of the time. Their deep complexes had been hand dug and were vast underground living quarters, complete with hospitals, kitchens, sleeping areas and ammo dumps. The tunnels were nearly impervious to most bombing attacks."

Approximately 2,000 meters to the east of Chu Moor Mountain, and across a stream laden valley, was the 1/22's temporary operating base and the fire support base (FSB) of C Battery, 4th Battalion, 42nd Field Artillery (4/42 FA) with their five, 105mm howitzers. The base was called LZ C-Rations.

A fire support base (FSB, firebase or FB) was a temporary military encampment widely used during the Vietnam War to provide artillery fire support to infantry operating in areas beyond the normal range of fire support from their own base camps. FSBs followed a number of plans, their shape and construction varying based on the terrain they occupied and the projected garrison.

A fire support base was originally a temporary firing base for artillery, although many evolved into more permanent bases. Their main components varied by size: a typical FSB usually had a battery of six 105 millimeter or 155mm howitzers, a platoon of engineers permanently on station for construction and maintenance projects, at least two landing pads for helicopters (a smaller VIP pad and at least one resupply pad), a Tactical Operations Center (TOC), an aid station staffed with medics, a communications bunker, and a company of infantry serving as the defense garrison. Large FSBs might also have two artillery batteries, and an infantry battalion. Under the original concept of the artillery fire support base, a six-gun battery set up with one howitzer in the center to fire illumination rounds during night attacks and served as the base's main registration gun. The other five howitzers were arranged around it in a "star" pattern. Smaller FSBs tended to vary greatly from this layout, with two to four howitzers of various calibers (usually 105mm and 155mm at battalion level) located in dispersed and fortified firing positions. These smaller bases arranged their guns in square or triangle patterns when possible.

LZ Swinger, another FSB and temporary home of the Golden Dragons of the 1st Battalion, 14 Infantry, was situated on a hill about 3,000 meters to the northeast of the Chu Moor Mountain complex.

Besides 1/22 and 1/14, several other units would be committed to the upcoming battle, including A/1/12, B/1/8 and several aviation, artillery, and engineering support units.In late April 1968, 1/22 and C Battery, 4/42 were relocated to the newly constructed LZ C-Rations. Roach remembered, "What we went to was the roughest site for a firebase I had ever seen. Engineers had to clear the trees with explosives and then a bulldozer was brought in by helicopter to push the trees aside."

Roach continues, "In the days the firebase was under construction air strikes were continuous. The jets would deliver their load and then put on a show for us grunts. They were so low the pilot's face was clearly visible. The planes came in low and went straight up and spiraling over the firebase. The dirt would be sucked off the ground. It was impressive."

"The LZ was on the lower slope. While we were setting up, units of 1/22 plus other units were air assaulted into the area around Chu Moor," continued Roach. As soldiers moved into the two firebases, company-size patrols were sent out in all directions to stir things up and see if the NVA had indeed made another incursion from their Cambodia sanctuaries. Company E/1/22 provided LZ security for the artillery batteries, which besides C/4/42 included C/5/16, C/5/22 and B & C/1/92 for a total of 25, 105mm artillery howitzer guns. The 5/16 and 1/92 Artillery were all 155mm howitzers and C/5/22 were 175mm and 8 inch howitzers.

On 20 April, Company C/1/22 patrolled the area north of LZ C-Rations and southwest of LZ Swinger. An unfortunate friendly mortar round landed near their location and wounded one soldier. This was an ominous beginning to what would become a tragic week ahead.

First Platoon Leader Lieutenant John Bobb (C/1/22) remembers: "About the third week of April, 1968, our company was choppered to the top of a hill, named 'Chu Ya Bruh.' I remember an EOD unit attempting to render safe an unexploded bomb. We were assigned part of the perimeter, and started our usual routine of preparing defensive positions including bunkers with overhead cover. This initial perimeter became the Battalion Command Post.

"I remember hiking out of the perimeter with the entire Company. I remember the hike because it was downhill for a long ways, with the heavy pack on my back, pulling me backwards, and my lower front quads burning from exhaustion. Evidently we patrolled at the base of a mountain for a few days.

"Usually the company travelled with three platoons; a headquarters element, the Company Commander, First Sergeant, and the company executive officer was usually in the rear. We had attachments; like medics from the Battalion HQ medical platoon, Artillery Lieutenant Forward Observer for the 105 battery, a 4.2 mortar FO for the battalion, E Co. 4.2 mortar platoon, and an 81mm mortar FO for the 81mm mortars, from E Co."

Sergeant Arlen Bliefernicht (B/1/22) recalls: "B Co. was assigned to firebase security, during the night we observed red and blue flashlights moving up and down Chu Moor Mountain. Our Battalion commander requested permission to shell the lights, permission denied. He then requested air strikes or gunships, permission denied. One of the men asked if he could snipe at the lights with a 50-Cal machine gun, single shot at a time. Battalion agreed to do so. I don't know if he hit any of the bad guys, but we let them know we knew about them."

Bliefernicht continues: "Next day, our battalion commander asked permission to recon Chu Moor Mountain. Permission denied. The battalion commander then said, they didn't say anything about the valley before Chu Moor Mountain. That was when D Co. was sent to recon the area."

In the early morning hours of 22 April, D/1/22 under the leadership of Captain James Burke, slipped away from their night position, just below LZ C-Rations, and moved off in a southwesterly direction, exploring the area below Chu Moor Mountain.

Accompanying the unit was Corporal Doug Stanek, who was the Acting Recon Sergeant and temporary RTO (radio telephone operator) for the FO (Forward Observer), 1st Lieutenant John Snellings. They were both part of the artillery battery (C/4/42) on the LZ, and were assigned to D Company to provide them with protective and offensive 105mm gun fire.

The 42nd Field Artillery has a long association with the 1/22 Infantry and the 4th Infantry Division. In Vietnam the 4th Battalion of the 42nd Field Artillery was a towed 105mm howitzer battalion assigned to the 4th Infantry Division, to render direct support to the 2nd Brigade, of which the 1/22 Infantry was a part.

In 1966 the 4/42 Field Artillery deployed to Vietnam on the same ship as the 1/22 Infantry.

Its personnel shared the same firebases as the Regulars and provided forward observers to 1/22 units in the field. Many Regulars are alive today, because of the fire support given to them by the Redlegs of 4/42. The 4/42 Field Artillery served in Vietnam from 1966-1970. The 42nd's motto is Festina Lente (Make haste slowly).

At about 0800, Captain Burke sent out an advance reconnaissance patrol to reconnoiter about 500 meters in front of his main force to act as an early warning should the enemy be sighted. The small patrol consisted of 1st Lieutenant John McKee; his RTO, Ronnie Colson; point man, Private Herbert 'Herbie' Hammond; PFC Tony DeVito; M-79 man, PFC Toby Van Skike; machine gunner, Specialist-4 Rainer Guensch; 4.2 Mortar FO, Specialist-4 Elbert Lynn Shadoan and his RTO, PFC Ray 'Skip' Diepenbrock. At 0900, Captain Burke ordered his company to saddle up and to move out in the same direction as the recon team.

The advance element soon came across several well concealed, freshly made, but empty bunkers. In a high state of alertness, the team waited for a short period of time and then quietly began their return march back to the company. "I'm still not sure how Herbie saw them.

Suddenly he opened fire and immediately received return fire from both left and right," remembered McKee. Diepenbrock reported, "Hammond had seen several NVA crossing our trail and opened up, killing one. We all hit the ground." The 4.2 mortar FO tried calling in mortars but it was denied as being too dangerous because the enemy was between the team and the company, despite the fact that Shadoan knew their exact position. He always kept a close eye on his map coordinates.

As Lieutenant McKee ordered his men forward again, point man Hammond was fatally wounded at 1345. Small arms fire immediately broke out on both sides. The NVA were observed wearing full packs and gear.

Guensch started to blaze away with his M-60 to provide protective cover for the rest of the team. "I fired that 60 so much that I had a cook-off and had to break the belt to stop it."

PFC Van Skike was not far behind and spotted movement ahead of him as the incoming bullets filled the air. He saw an enemy soldier hiding in a hole with his left arm raised to throw a grenade. He immediately shot his M-79 in the direction of the NVA and killed him before he could toss the deadly explosive.

"I was just a few yards behind Hammond and moved up under fire and dragged him back about 20 yards," recalled PFC DeVito "Lieutenant McKee then hoisted Hammond, then still alive, on his shoulders and Guensch, Diepenbrock, and myself provided cover fire while the team retreated." Tony DeVito would later be awarded with the Army Commendation Medal with V device for this action. Rainer Guensch earned the Bronze Star for his bravery under fire during the fighting.

As Colson called in a medevac on the move, the team headed back to an LZ position about 350 meters up the trail where they set up a small perimeter. At first, the evacuation helicopter was reticent to pick up Hammond as the enemy was all over the area and it had difficulty in seeing and identifying friend from foe. A quick thinking Colson pulled out his machete and chopped out a spot in a bamboo clump. He then waved his Confederate flag that he had brought from Georgia and finally convinced them to move in and lower their basket. Hammond's last words were, "Somebody, please help me breathe."

According to PFC Diepenbrock, "Guensch was on the trail, De-Vito to the right and I was on the left front and the rest of the patrol was behind us." Several minutes later, seven more NVA soldiers silently moved around their position and were just 20-feet away to their rear. Guensch heard something and turned his head in the direction of the sound. "As soon as I took a quick glance behind me, I swung my machine gun around and opened up on the seven soldiers before they began to fire back." Three of the NVA went down. The NVA returned fire and shot Guensch's M-60 right out of his hands. "That's when I saw they shot one of my bipods off, smashed my forward hand guard and put a round into my receiver. I couldn't get it to work after that. All I had now was my .45 pistol with just three rounds."

Lieutenant McKee continued, "The enemy had thrown a grenade that slightly wounded Guensch and his assistant gunner when I ordered the patrol to fire in all directions."

Diepenbrock stated, "I slid around to the side of a tree to see if any enemy were coming on to me on my left side and spotted three NVA soldiers approximately 20 feet away shooting at the M-60 gunner on my right. I opened up on them with a short burst and killed them."

The patrol quickly bolted further back along their original route, calling in an effective gunship run of rockets and mini-gun fire on smoke grenades that were left behind to cover their retreat.

The team was eventually able to link up with the rest of their company. Lieutenant McKee would later state, "To this day, I think of Herbie, and the fact that, but for his observant eye, I might not be saying this, and the remaining elements of that patrol might also have our names displayed on The Wall." (The Vietnam Veterans Memorial, Washington, D.C.)

At 1540, D Company continued to back track its way through the valley and eventually established a nighttime position.

The next morning, 23 April, the Company headed back to LZ

C-Rations where they regrouped, re-supplied, and rested up for what would soon be one of the biggest fights of their lives.

PFC Roger Salber had just arrived in country and had been assigned to D Company 1/22. He met up with his Battalion on the LZ on 23 April. The next day, he would be tested in battle for the first time.

CHAPTER 5
Day 1: Wednesday 24 April 1968

"The unseen enemy had deep holes in the ground, called spider holes. These holes were called this after the trap door spiders that jumped out and captured its prey. The enemy soldiers slowly opened their camouflaged lids and then fired their deadly B-40 rockets at us before going back into hiding under the ground to wait for another opportunity."

DAILY STAFF JOURNAL OR DUTY OFFICER'S LOG

ORGANIZATION OR INSTALLATION		S-3, 1st Bn 22nd Infantry REGULARS
LOCATION		FSB YA868888
PERIOD COVERED		0001 24 Apr 68 - 2400 24 Apr 68
ITEM NO.	TIME	INCIDENTS, MESSAGES, ORDERS, ETC.
10	0725	Airstrike today: (1) YA83884 at 0930 hrs. (2) YA8584 at 1300 hrs.
18	0844	D Co location YA858875, K-9 alerted, something east of river.
19	0845	Co D reconning by fire as the dogs were alerted to something east of river.
35	1140	All elements of A Co have closed PB. 3rd platoon found a fresh trail at YA860913. The trail runs NE along blue line in this area. Trail is just a few hours old. Tracker team says that 10-15 NVA using trail.

36	1200	The trail found by A Co 3rd platoon is approximately 2 ft. wide & has trees overhead hiding it.
42	1314	Airstrike completed at this time.
43	1320	C Co point has movement to the front.
53	1505	B Co 2nd & 3rd platoons & Hdq checked out of the assigned AO given to them by Bn 3. Found 1 trail running to SE. Will work way NE to FSB
56	1529	D Co has reached top of objective, discovered 6 NVA, and took enemy under fire, enemy fled.
57	1534	D Co present location YA844891—enemy closing from West & South.
58	1537	D Co in contact at this time w/unknown size enemy size force.
59	1539	Bn 3 put in request for gunships.
62	1544	D Co believes that they are in contact with large force. Doesn't know what size yet.
63	1546	D Co requested help fast—Bde trying to get guns as soon as possible.
64	1548	Guns enroute this time, are Cougar 13.
67	1602	D Co requesting basic load of ammo.
68	1611	D Co has 1 WIA.
69	1613	D Co needs ammo asap.
72	1623	Gladiator 063 was hit, lost transmission oil pressure.
73	1625	Gladiator 063 crashed at FSB, requesting dustoff from our A/C channel.
79	1701	So far D Co has 5 WIA, 4 not serious. Situation—sniper fire and most of the enemy have moved to the west in a valley.
80	1704	Enemy was, they think, a company size unit.
81	1712	D Co in heavy contact at this time.
82	1715	D Co has 6 WIA—2 for dustoff.
83	1717	D Co contact is light now.
84	1730	Dustoff D Co 1 WIA completed. Need 1 more.
85	1736	D Co has 7 WIA of which 4 are for dustoff.
86	1737	D Co contact has quieted down. NVA fled west & NW. Did spot some to the west, about Co size.

88	1758	Wrap up of D Co contact: at YA844891 1st contact was with 6 people—fled to west, came back with larger force, we put in arty & gunships. D Co made contact on top of hill, are now on top of hill in perimeter. Had contact with Co size force, enemy fled W to NW, was a number of exchanges of fire, heavy & light. Results: 7 WIA, US, of which 4 needed dustoff. Negative results as far as can tell on enemy.
89	1822	Dustoff of 3 remaining litter cases completed at this time.
90	1825	D Co receiving sniper fire at this time from the SW. D Co confirming 2 NVA KIA.
94	1853	D Co situation report: everything quiet; getting positions fortified; breaking up ammo; checking out rest of company.
96	2115	Final count of D Co WIA's, 10 WIA, 4 were dusted off today and 6 will be dusted off tomorrow. 3 wounds were serious; the remaining 7 were minor but need Dr attention.
101	2320	Request Snoopy mission tomorrow within the following grids: TA7692/YA9092 TA7687/YA8087.
102	2345	Snoopy mission turned down.
104		Summary: 1/22 continued search and destroy operations in assigned AO with heavy contact with enemy forces. D/1/22 had contact at YA844891 with 6 enemy. Then enemy hit with large forces. Gunships, Arty and Airstrikes were employed. Results: 2 NVA KIA and 10 US WIA. Area of contact will be swept tomorrow morning for enemy bodies and equipment.

In the early morning hours of 24 April 68, Companies A, B, C, and D of the 22nd Infantry Regiment moved out in separate directions with their ultimate goal of reaching the crest of Chu Moor Mountain to determine the extent of the enemy's presence. A Company took off towards the north, B to the east, and C and D to the southwest. D Company planned to skirt around the south end of the mountain to make its ascent approach from the west. At 0845, a scout dog leading D Company was alerted by possible enemy movement across a river and began barking.

Some dogs are trained to silently locate booby traps and concealed enemies such as snipers. The dog's keen senses of smell and hearing would make them far more effective at detecting these dangers than humans. The best scout dogs are described as having a disposition intermediate to docile. They are not as aggressive as attack dogs.

Scout dogs were used in Vietnam by the United States to detect ambushes, weapon caches, or enemy fighters hiding underwater, with only reed breathing straws showing above the waterline. The US operated a number of scout dog platoons (assigned on a handler-and-dog team basis to individual patrols) and had a dedicated dog training school in Fort Benning, Georgia.

The Scout Dog Team would join a troop unit—generally an infantry unit—and would walk "point" ahead of the other troops, searching for potential signs of danger and peril, such as trip wires, snipers, mines, or potential ambushes, as well as for potential aids such as hidden caches of food or weapons.

The dog, who could travel relatively obscured by the underbrush and unseen by any enemies, would know to signal his handler as to the presence of any of these items of interest, and the handler could likewise signal to the patrol leader, letting him know whether or not to move ahead, and potentially saving the lives of all the troops involved in the mission.

After the scout dog with D Company alerted, the Company then began an advance reconnaissance by fire but ultimately received nothing in return, so they continued on their march up the mountain.

To the north, A Company located a fresh, two-foot wide trail that was covered by overhead trees. It was believed to be an NVA re-supply route as it headed northeast along a stream. The trail was checked out and it was determined that 10-15 NVA had been using it in the recent past.

Minus his broken machine gun, Rainier Guensch started up the hill with D Company with just his .45 pistol. He was able to borrow a few grenades from some of the guys, but nothing else. He was told he would get a new M-60 during a scheduled re-supply mission later in the afternoon.

At 1505, B Company also located a trail running southeast and was ordered to work its way back northeast to LZ C-Rations. The other companies continued on towards their day's objectives. Specialist-4 Mike MacCallum, an RTO for the 1st Platoon Leader in D Company, described the day's conditions, "The valley between C-Rations and Chu Moor was very deep and it was hot. April is the hottest month in Vietnam because it is the last dry month before the monsoons start. We were heavily loaded, carrying about 65 to 85 pounds per man.

"The mountain was high and the side we climbed was mostly covered with grass, shrubs, and a few trees here and there. At times it was so steep that we had to grab the bushes and pull ourselves up by hand. We ascended in single file. The rotational order of march for the day was 2nd Platoon on point, followed by the 3rd Platoon, and my 1st Platoon brought up the rear."

Stanek and Lieutenant Snellings (C/4/42) working with D Co, 1/22) knew that they needed to be constantly aware of their position. From time to time they called in smoke rounds to assure their exact location. They also realized that each of their 105mm howitzers was capable of firing 33-pound projectiles of high explosives nearly ten miles away and that they had a killing diameter of 50 meters, sending out hot steel fragments in all directions upon impact. During the morning hours, preparatory artillery rounds had been hitting the top of Chu Moor Mountain in advance of the 1/22 companies. The enemy now knew for sure that the Americans were moving in on their well dug in positions. The American soldiers hoped the fire would clear the way for their safety.

Lieutenant John Bobb (C/1/22) remembers, "One day we were given orders to approach a mountain named 'Chu Moor', along a ridge line. I do not remember what I knew at that time; however have learned since that Delta Company had stepped into a superior enemy force on another ridge line leading up the same mountain.

"Normally, at that time an infantry platoon was supposed to have 44 members, and the rifle company probably should have had four platoons, and add-ons to total over 200 men. But now the platoons normally had about 18 to 30 men.

"Our C Company, was going up the opposite ridge line, sensing increased heavy enemy movement to our front. We were alone on our ridge line. A and B companies were dispatched to go up behind D Co. on their ridgeline, so they were no longer trapped from being able to maneuver freely."

Specialist-4 Elmer Hale was a rifleman with C Company. Partially up Chu Moor, Hale stood by a tree talking to Staff Sergeant Hollis 'Pappy' Buck. "Buck was a career man and we got to be very close, which can be the wrong thing to do in combat. He was 38-years old and had fought at the Chosin Reservoir during the Korean War."

"'Pappy, something ain't right!' 'What do you mean?' he said. 'There's some shit up here somewhere. We're going to hit some shit. I can just feel it.' The monkeys had quit chattering and the birds quit singing. It was real quiet and the jungle was very thick."

C Company's commanding officer, Captain Leo (Larry) Konnerman, sent a small recon team up ahead and they soon observed movement in the brush off to their right front and reported it in at 1320. Captain Konnerman now decided to split up into four platoons, each led by a lieutenant.

1st Platoon leader, Lieutenant John Bobb, spread his soldiers out and advanced without reconnaissance by fire. He did not want his men to disclose their position in the thick underbrush. If one of the platoons made contact, the others would be able to quietly outflank the enemy.

"The guy in front of me let a tree branch fly into my face, breaking my eye glasses," recalled C Company medic, Charles Shyab. "Along with getting smacked in the face, the branch knocked my helmet off and it rolled down the side of the mountain without any chance of recovering it. So for some kind of eye protection, I put on my old scratched up sunglasses."

Over in D Company, Corporal Doug Stanek stated, "Our dog-tired company desperately needed to rest and be re-supplied with food, water, clean uniforms, and perhaps a happy letter from home. The trail was now wet and slippery from a rain, making the hard climb even more difficuLieutenant" Tony DeVito recalled, "Some of the men were making jokes like there was going to be hot chow, mail, and a reenlistment officer on top of the mountain when we reached it."

Specialist-4 MacCallum remembered, "As we climbed the mountain in the heat, the German shepherd (scout dog) began to tire. At one point the entire company stopped to let the dog rest. When the dog needed to rest again, the company commander refused to stop, instead the dog and his handler stopped and let some of the company pass by. Eventually, the handler ended up all the way back at the 1st Platoon. The handler carried the dog over his shoulders for a while, but the dog was big and that didn't last long. When the point reached the top of Chu Moor, the 1st Platoon may have become slightly separated from the rest of the company because of the dog."

As was typical with NVA strategy, they let the advanced elements of D Company move forward into a waiting trap, hiding in their fighting holes and in the thick jungle. They wanted the Americans to get close to them so American artillery and air power could not be used effectively without endangering their own friendly forces.

At 1529, the leading elements of D Company reached the top of their objective and ran into six NVA. PFC Roger Salber related, "My Platoon Sergeant and myself were two of the first to make contact and were pinned down by machine gun fire. I took a shot through the seat of my pants as I was lying there trying to figure out which direction the shots were coming from. We were in thick jungle undergrowth. Some well-placed grenades from the platoon leader (Lieutenant Hugh Crumple) got me out of this mess."

"We were stuck," reported Stanek, "and the enemy now knew our size and positions. We made a huge mistake by walking right into the enemy's hidden bunker lines, dug deep into the side of Chu Moor Mountain."

The single file company became separated into two groups during the initial outburst of fire and many soldiers simply dropped their rucksacks and made an attempt to link up with the forward elements. Initially the enemy fled but within minutes they returned, bringing with them an NVA company. MacCallum stated, "From my position down below the brow of the summit, I could hear rifle fire and explosions ahead. It was very intense."

Specialist-4 John Piraino was one of the soldiers who had dumped his rucksack. He quickly hit the ground near the trail just as a rain began to fall. To his amazement, right out in front of him, several NVA soldiers appeared out of nowhere. Piraino sank lower into the brush and allowed the lead enemy soldier to approach closer to his position. He could not believe the situation that he was in. As the NVA knelt down on one knee next to a fallen tree, he quietly signaled for his men to follow him. Piraino glanced around and could not see any more soldiers but he could hear them as they came closer and closer. He silently moved his M-16 selector switch on to full automatic and let go with a burst of a dozen rounds at the kneeling soldier, hitting him in the head and chest and killing him instantly.

Specialist-4 Wayne Raymond, a trained infantry scout, quickly determined that part of the company was split off and took PFC Bruce Gass and PFC Toby Van Skike and snuck back down towards the blocking enemy force. They came across the hidden NVA and opened up with M-16s, grenades, and Van Skike's M-79 grenade launcher, surprising and inflicting many casualties on the enemy. Their successful ambush allowed more of the company to link up and form a perimeter on top of the mountain.

Specialist-4 Rainer Guensch, still without a workable machine gun and armed only with a .45 pistol and three rounds, heard the firing up the trail. He thought to himself, "What kind of shadows are those stupid assholes firing at now." When he finally neared the top, he began to see dead enemy soldiers scattered about the ground and realized that this was the real thing. As quickly as he could, Guensch collected the dead enemy's weapons and ammo, so he would have some kind of protection.

Corporal Doug Stanek recalled, "A close friend of mine, Pete Becker, the D Company RTO, was lying on the jungle floor as I arrived at the ambush site and dove right next to him. I was completely out of breath. There were enemy bullets and B-40 rockets coming at our position from every possible direction. We could not even hear ourselves think, let alone talk into our radios. I searched for the officer in charge and saw through the smoke the Company Commander, Captain James Burke, and my FO, Lieutenant John Snellings, about thirty feet away. The CO needed Becker's radio to call in gunships and the FO needed mine to bring in artillery. Captain Burke waved at both of us to move over under the heavy enemy fire. I screamed as loud as I could to Pete that we had to move out now! Pete yelled back that he had been shot in both legs." Another soldier saw their predicament and helped Stanek move the wounded RTO to his Captain. Stanek had taken his radio and rucksack off during the action and now he had to go back under fire to recover his radio. He held his breath, dropped his M-16 on the ground and ran like hell to recover the radio and drag it back to his FO.

Cobra gunships moved in and began to fire dozens of high explosive rockets into the jungle. They came in fast and low, raising all kinds of hell. On their second pass, they unleashed thousands of rounds of M-60 machine gun ammo.

A second request for gunships and artillery immediately went out. The FO team of Corporal Stanek and 1st Lieutenant Snellings called in 105mm rounds from the supporting fire bases and did their best to help hold off the enemy attack. Specialist-4 Dennis Scholar and the 1st Platoon's weapons squad leader, Specialist-4 Jim Plumb, were located just off the trail where it came up the mountain, not far from the FO team.

"Plumb was immediately hit with shrapnel and wounded in the belly," stated Scholar. "Jim fell to the ground screaming." Stanek continued, "Plumb had been hit by my third salvo of 105mm rounds. We didn't really know where our own men were on the mountainside." The enemy had closed to within twenty meters in some areas of the perimeter. "I remembered shouting into the radio (PRC-25) to stop the artillery, which must have been heard because that was the only salvo that came in on us," stated MacCallum.

PFC Van Skike stated, "I heard the Redleg FO call for a cease fire on fire support. Unfortunately the rounds kept coming in. Again I heard the FO call cease fire and the radio crackled and the reply came back, we haven't fired a shot in five minutes!"

Companies A and C were told to hold in place on the hillside and watch for possible enemy movement in their operational areas. Doc Shyab with C Company stated, "It was in the afternoon and we began digging in for the night. One of our well hidden observation posts spotted some North Vietnamese soldiers going up the trail that we were near."

Within a half hour Company D was nearly out of ammunition and two requests for a re-supply went out. First, "D Company needs a basic load of ammo," and then a more urgent call, "D Company needs ammo ASAP!"

Captain Burke then asked for volunteers to go back down the trail to the dropped rucksacks and gather any ammo and food left behind. John Piraino, Specialist-4 Joe Salvatore, and Specialist-4 Raymond accepted the challenge. Several minutes later, they successfully returned with sandbags full of ammo and C-Rations to be distributed among the men.

At about 1620, helicopters moved in over the perimeter and began dropping crates of ammo as it was much too dangerous to land. Specialist-4 MacCallum remembered, "The wash of the blades of the helicopter that hovered above us pushed the leaves and branches around so you could catch glimpses of men in the helicopter throwing crates of ammo out the doors as fast as they could. I have always admired the bravery of those guys."

One of the helicopters, Gladiator 063, took seven or eight bullet hits and lost its transmission oil pressure. Somehow, it made its way back to LZ C-Rations where it purposely crashed just outside the firebase, without inflicting any injuries to the crew or the soldiers on the base.

Bliefernicht of B/1/22 recalls, "Late that afternoon we heard D Co. was hit hard while trying to set up a night location. When choppers were approaching D Co.'s L.Z., the slicks were drawing heavy sniper fire from the trees. One such slick was shot up and was going to try to make it to our firebase. Our bunker was by the L.Z. and I watched as the slick came in. The chopper started to sway back and forth. I later learned that the hydraulics was shot out. The chopper crashed by our L.Z. and near our bunker. The rotor from the tail of the chopper bounced across the L Z. and landed in front of our bunker. Only minor injuries were suffered by the chopper crew."

Assigned as the LZ medic, Specialist-4 Roach stated, "Those of us at the firebase kept up with the developing battle by hearing the radio transmissions. We could see air strikes and helicopter gunships as well as artillery barrages from our site. We knew it was a major operation."

Back on Chu Moor, MacCallum recalled, "There was a small, reasonable level clearing near our position below the top of the hill. Our platoon leader instructed one squad and me to secure the clearing for a landing zone and he called in for a dust off for Plumb. Then he took the rest of the platoon up to the summit to join the company, which I estimated to be about 50 meters away."

MacCallum continued, "The helicopter came in low, rising up the mountain, so I think it was pretty well protected. I thanked the pilot with all of my heart for coming into that situation to pick up our wounded. Once the dust off was over, we climbed our way to the top of the mountain to rejoin the company. We found the company had formed a perimeter. The 1st Platoon occupied the third of the perimeter where we had come up the mountain, the furthest from the enemy positions and were digging in. The ground was hard, there were periodic B-40 rockets and rifle fire incoming and we had to stay low, so progress was slow. It was an absolute misery to be so scared and to feel so hopeless. As I recall, we were on a pretty flat ridgeline that must have led to the actual summit of Chu Moor. We were on the brow of the ridge and the mountain slopped away below us. Our bunker (with the platoon leader and medic) was actually slightly above the line bunkers on the perimeter. The ridge was heavily forested and it was difficult to see very far in any direction. Above, the leaves of trees formed a nearly solid canopy over our heads."

Once the injured Plumb was removed, Dennis Scholar immediately took Jim's machine gun and all the M-60 linked ammo he could carry and crawled under heavy fire up near to Captain Burke's position. He then helped set up the gun and the weapon's team commenced firing. Enemy B-40 rockets and automatic rifle fire blazed over the heads of the stranded company as it tried to dig in and return fire to the best of its ability.

As the battle continued, D Company fortified their positions with sand bags, digging in deeper and using chopped down trees to better their protection. Specialist-4 John Piraino, with the 1st Platoon, was situated in the middle of the perimeter trying to dig in. He had filled several sandbags, and piled them up behind him in the direction of the enemy. As Piraino swung down with his entrenching tool, in order to dig his hole deeper, a 122 mm rocket round struck the pile of sandbags. Fortunately, he was in the bent over position and the blast was stopped from cutting him in half. However, his luck would not last for long.

"I had just gotten three layers of sandbags filled for my protection in front of my bunker when the K-9 scout dog that was with us joined me in my hole. I was petting the dog while I was on the lookout and admiring his courage. Suddenly I heard a loud whistle and then a boom! Joe Salvatore, a good friend, who was near me heard me say, 'I am hit, I have been hit.' Joe looked out of his hole and saw my face and arms covered in blood and figured I was a goner for sure. He yelled for a medic who patched me up as best he could. The dog that was with me was also wounded very badly and was bleeding and moaning in pain."

Once Piraino was helped down to the evacuation zone, the dog handler came up to him and asked if his dog could take Piraino's place on the last medevac chopper of the day. The handler begged him to help save the life of the dog. John agreed, being the man he was, and returned back up to the perimeter where his life would continue to be in danger once more.

At the Company D perimeter on Chu Moor, the firefight slowly died down as the NVA fled west and northwest at 1737. Occasional sniper fire continued throughout the night as the soldiers fortified their defensive positions. Ten men had been wounded, three seriously, and four were cabled up to medevac helicopters in the evening. There were only two confirmed NVA killed in action (KIA) but a search was planned for the morning to look for more casualties.

In the 4th Infantry Division Newspaper (Ivy Leaf), dated 19 May 1968, Captain Burke was quoted as saying, "The NVA soldiers were in trees, in spider holes, and just about everywhere we looked."

"The unseen enemy had deep holes in the ground called spider holes," recalled Corporal Stanek. "These holes were called this after the trap door spiders that jumped out and captured its prey. The enemy soldiers slowly opened their camouflaged lids and then fired their deadly B-40 rockets at us before going back into hiding under the ground to wait for another opportunity."

As darkness settled in over the Chu Moor Valley, it was decided that B and C Companies of the 1st Battalion 14th Infantry would move out in the morning and sweep the high ground west of LZ Swinger where there were indications that an enemy supply point might be located.

"I honestly don't remember much about that first night up on Chu Moor," stated MacCallum. "I don't remember eating or being short on water. I don't remember how we went to the bathroom, pinned down like we were, but we must have found a way. We must have kept watch and slept as best we could when we were off watch. I remember artillery being called in after dark and flares so that we could spot the enemy if they attacked."

By 2300, the FO team learned that C Battery, 4/42 located on LZ C-Rations had run out of 105mm ammo. Fortunately, the enemy did not realize that there was no longer any artillery protection available. Men, such as John Piraino, continued throwing grenades and rocks (to fake out the enemy) throughout the evening hours. A call for Spooky (Puff the Magic Dragon) gunships was turned down at 2320.

Stanek recalled later that night, "I leaned against a tree and used it as a backrest. Wet from the rain and completely exhausted physically and mentally from the intensive attacks, I passed out almost instantly with my hands holding my loaded rifle on my lap. A man could not tell if he slept for twenty minutes or two hours up there. There is no such thing as time in the reality of battle. Fortunately, during the early morning hours of day two, C Battery received hundreds of rounds of 105mm ammo brought in by Chinook helicopters."

Stanek continued, saying, "There were enemy soldiers running around us like rats in the black jungle. Enemy snipers were tied securely to their trees so they didn't fall while they slept during the night. We heard NVA movement all around us but we could not see one soulless killer through the darkness."

Lieutenant Bill French, Medical Platoon Leader recalls, "What I do remember from my vantage point in the firebase was witnessing the air strikes and the artillery impacting on Chu Moor Mountain. What started out as green mountains soon turned into charred and shattered landscape. Through the artillery spotting scopes we could actually see snipers in the trees. Fire directed by the enemy towards the gun ships and airplanes was evident and more than one gunship landed in the Firebase after being hit. The typical injury was the neck and face of the pilots as the chest plates protected them somewhat in the torso. The real excitement was boarding the choppers to treat and remove the air crews while the machine guns were still armed and smoking.

"I did spend some time in the battalion command post to try to learn what was going on and what supplies would be needed by the medics. My most vivid memory is how the symbols of the companies were moved around on the map to counter the suspected moves of the enemy. I often wonder if those who were moving these symbols around knew what it would take by the soldiers on the ground to move from point A to point B considering the terrain, heat, heavy packs, etc. I do not recall that many of the officers in the command post had actually been company commanders in Vietnam.

"I do remember frustrations in getting the right supplies to the medics. The system available was not near as efficient as for guns, ammo, and food. It was often necessary for the medics at the Brigade Trains and Base Camp to beg, borrow, and steal to get what we needed. I do remember one helicopter specifically carrying medical supplies, losing much of what was in the net before it reached the company."

During the first day on the mountain, the Army utilized one gunship, two air strikes, 1,023 artillery rounds, and 177 mortar rounds to fight off the enemy's attack. These numbers would greatly increase on the following day.

DAILY STAFF JOURNAL OR DUTY OFFICER'S LOG

ORGANIZATION OR INSTALLATION		S-3, 1st Bn 22nd Infantry REGULARS
LOCATION		FSB YA868888
PERIOD COVERED		0001 24 Apr 68 - 2400 24 Apr 68
Item No.	Time	Incidents, Messages, Orders, etc.
	1630	D/1/22 has enemy contact. Initial 6 NVA now an undetermined force. Mike 84 requests gunships and to have a AS on standby. Gladiator 063 has gone down in the contact area.
	1645	Gladiator 063 brought down by 7 or 8 rds. of enemy small arms fire. Bird completely destroyed. Pilots and gunners O.K. D/1/22 has 1 US WIA.
	1700	D/1/22 has 5 US WIA's. 4 not serious, 1 serious. Still receiving sporadic sniper fire. Most of the enemy fled to the W. Believed to be a company sized element.
	1720	D/1/22's contact considered light. Report 1 more US serious WIA. Enemy continues to flee to the west. Summary GOLDEN DRAGONS continued operations in assigned AO. Alpha Co while sweeping to the north found a large concealed trail network at grid 857579. Negative enemy sighting. Delta Co 1/22 had contact south of the Bn AO. Enemy fled to the west. PLANS: Bravo and Charlie Co's to move west and sweep high ground west of LZ SWINGER. Where indications are that enemy supply point might be located. Delta Co to assume responsibility for firebase security.

4TH INFANTRY DIVISION
OPERATIONS SUMMARY
21 APR 68—30 APR 68

24 April. At YA932927 Company D, 2nd Battalion, 35th Infantry found one NVA body and another at YA930926. Both bodies had been dead for about two weeks. The aerial rifle platoon of Troop D, 7th Squadron, and 17th Air Cavalry was inserted at YA745949 to check out a bunker complex sighted by Troop B. The platoon conducted a reconnaissance with negative findings and began extraction when six US were wounded by SA and AW fire from the east of the PZ. At YB930825 two MSF companies engaged an estimated four NVA platoons, resulting in one US WIA, one Australian MIA, two MSF KIA and five MSF WIA. At YA933930 Company B, 2nd Battalion, 35th Infantry found five SA weapons. At YA783427 LRP 4B engaged several NVA, killing one and capturing an AK-47. At YA844891 Company D, 1st Battalion, 22nd Infantry made contact with an estimated NVA company. The fighting, which lasted for two hours, resulted in seven US WIA and two NVA KIA. A UH-1H helicopter in support of the contact was shot down by ground fire, but all personnel were rescued with only slight injury to one US.

CHAPTER 6
Day 2: Thursday, 25 April 1968

"The close in air support was like in the movies. You could hear the 'pop' on the aircraft when they dropped the bombs and hear the trees hit by the wings."

DAILY STAFF JOURNAL OR DUTY OFFICER'S LOG

ORGANIZATION OR INSTALLATION		S-3, 1st Bn 22nd Infantry REGULARS
LOCATION		FSB YA868888
PERIOD COVERED		0001 25 Apr 68 - 2400 25 Apr 68
Item No.	Time	Incidents, Messages, Orders, etc.
12	0735	D Co has movement to NW, are going to recon by fire.
14	0800	A Co 1st Platoon found NVA helmet with markings inside, LAI 1/9
16	0806	D Co just shot 2 snipers out of the trees.
19	0844	D Co has snipers around him, not a large unit.
20	0846	D Co has contact with enemy snipers.
22	0905	Red 16 just received fire from the west.
23	0912	D Co had 2 explosions from his south. Negative rockets from guns.
24	0915	C Co reports his FO hears sounds coming from north to northeast.
25	0920	D Co reports mortars have stopped and are receiving sniper fire at this time.

26	0923	D Co reports he believes the mortars are 60mm and are coming from the SW.
27	0926	D Co CO reports that he is in heavy contact at this time.
30	0937	D Co CO requests a basic load of ammo.
35	0949	D Co needs dustoff for 1 man with leg almost cut off, also needs ammo.
37	1003	D Co has 2 WIA—1 real serious.
39	1009	D Co has contact with company size force.
40	1010	Gator 524 has been hit, has 1 WIA crew member, and has neg oil pressure 15/more hits.
41	1015	A Co 3rd platoon hears what he thinks are outgoing mortars at location YA870909, heard 4/5 rds.
47	1053	D Co has sniper to NW & N side—about 2 of them.
50	1106	D Co has 4 WIA today & 7 confirmed NVA KIA's.
51	1112	D Co contact started again.
52	1118	Airstrike over, Tornado White going in.
53	1120	Had 1 secondary explosion during airstrike.
55	1122	Cider 30 received ground fire from sniper, YA838891, using 20mm now.
66	1213	Airstrike is completed at this time.
67	1215	D Co reports receiving incoming mortars as the airstrike was being put in.
70	1229	Tornado 21 finished at this time. Gambler 9 is going in.
73	1236	Gambler 9 finished at this time. Will start Redleg at this time.
79	1322	D Co has 23 WIA; 1st platoon-8, 2nd platoon-11, 3rd platoon-4, Hdq platoon-0.
80	1332	D Co has 1 NVA KIA, 8 all total.
82	1337	D Co FO just got wounded, registered fire.
84	1340	D Co has total of 25 WIA.
86	1350	A Co location YA858909, this location found fresh bunkers 3-4 days old.
87	1352	Airstrike going in at this time.
92	1418	D Co has 24 WIA / 6 of which are serious.
95	1421	D Co CO informs that he has contact from NW and has received 2 B-40 rockets from NW.
98	1431	D Co has 1 KIA & 25 WIA.

100	1440	D Co informed us that enemy size is Co or larger.
107	1517	Quiet at D Co location at this time.
108	1519	Dustoff 34 got the worst wounded out, 1 WIA dustoff now.
109	1530	D Co at this time has another KIA, US.
115	1618	Dustoff of 4 WIA (not serious) completed & Tornado guns departed D Co location. Will be back in 30 minutes to pick up the remaining 3 WIA's.
119	1640	D Co has 2 additional WIA's that have to be lifted out which will make a total of 5.
121	1650	D Co reports another B-40 rocket incoming at this time, from west.
124	1700	C Co has movement to their front at this time.
126	1704	C Co is pulling back into perimeter and waiting on prep of Redleg to his front.
129	1723	D Co has dusted off 7 and still has 3 to go.
134	1755	D Co 3rd [platoon on their LZ are receiving shrapnel from airstrike.
135	1757	D Co at this time has 2 NVA left to be dusted off. Has 1 of them on pad & 1 KIA to be taken out.
136	1802	D Co has had 9 WIA's dusted off.
137	1803	C Co CO says he hears mortar tubes popping at this time and is shooting azimuth to it at this time.
138	1808	Airstrike completed at this time.
143		C Co CO says enemy is in saddle of coordinates YA 843886.
145	1847	D Co casualties as follows: 1st platoon 10 WIA's, 2 KIA's; 2nd platoon 13 WIA's, 0 KIA, 3rd platoon 5 WIA's, 0 KIA; Hdq platoon 2 WIA's, 0 KIA. Still has 13 to be lifted.
147	1910	Airstrike completed at this time.
150	1950	D Co 3rd platoon reports last WIA lifted.
151	2005	Spooky 21 was requested by Dustoff 34 at 1845 hrs. Spooky is now working for this unit.
155	2039	D Co confirmed NVA KIA count at 2 yesterday.

| 157 | 2046 | D Co CO reports the following information on enemy element: Among the personal gear on enemy bodies were found some new uniforms, some small bags and 2-3 letter diaries, handkerchiefs and scarfs, possibly from wives or sweethearts from home. For headgear most of them were wearing camouflaged hats and carrying potato masher grenades. The unit was observed, yesterday, carrying heavy packs but today they had no packs. Captured were estimated: 1 SKS, 2 AK-47's, 1 RPD machine gun, 2 magazines. Also, found brand new AK-47 ammo clips, some still in wrappings. A pair of green goggles, no gas masks. All foot gear was HCM sandals. 1 KIA yesterday did not fit the description of a Vietnamese or Montagnard. He was very yellowed skin and smooth complexioned about 5'3", 120 lbs., and 18-19 years old. D Co also reports today had about 6 B-40 rockets incoming. |
| 158 | 2126 | 30 casualties were dusted off. 8 serious. |

Day two, 25 April, on the mountain found D Company initially involved with sporadic sniper fire and was able to kill two of them hiding in trees. "We did a sweep in front of our position and then took a break which didn't last very long," stated Roger Salber.

"The first incoming shot on the second day, just missed my head and hit my squad leader in the chest as we were talking. He desperately crawled to the center of the perimeter. I could not help him as we were under such heavy fire at this time. The incoming bullets just never seemed to end. They were even coming from above, out of the tree tops. I was dug in with sand bags between two very large trees. At one point I noticed the bark was stripped away from the trees and the sand bags were getting lower, as the dirt ran out the front."

By 0930 contact again became heavy with a reinforced NVA company attacking with mortars, rockets, and small arms fire. An intense enemy ground assault hit directly at the 1st Platoon that was just barely beaten back.

"I was trying to deepen my pitiful little foxhole when everyone around me hit the dirt and I saw a B-40 rocket on the ground right before me," recalled Rainer Guensch. "I quickly grabbed it and threw it away. Thank God it was a dud this time!"

"The 1st Platoon took about half of our casualties on the mountain that morning," recalled Mike MacCallum. "From the platoon leader's bunker, I was away from the perimeter and not engaged in the fighting against the assauLieutenant I monitored the radio for the platoon leader. I remembered that the amount of incoming fire, both small arms and mortar and rocket fire was incredible. It was just a constant roar of incoming and outgoing fire."

Once again, a call went out for more ammo, which stressed the severity of the ongoing battle. Another re-supply helicopter, Gator 524, took an incredible 15 hits, one wounding a crew member, and it too, barely made it back to the LZ before crashing safely. The day continued with on and off fighting.

PFC Salber stated, "Sergeant Vernoy crawled into my foxhole and we were throwing grenades, directed by someone to our rear. I do not remember if it was Lieutenant Hugh Crumpler or not. I remember clearly someone yelling, 'more to your right.' There was no way we could lift our heads above the sandbags! I had been elected to carry the bag of grenades and shovel up the hill the day before." In the 19 May 1968, *Ivy Leaf* article, Sergeant Steve Vernoy later stated, "It was rough. At one point there were so many NVA soldiers that I just kept chucking grenades until I ran out. I got six I know of."

During the NVA attack, PFC Tony DeVito volunteered to carry injured soldiers down to an unsecured landing zone, under a heavy volume of enemy fire, about 200 meters away. For his performance of this hazardous task, DeVito was awarded his second Army Commendation Medal of the Chu Moor Operation. DeVito also remembered, "There was an NVA soldier sitting right in front of me leaning up against a tree and staring at me the whole time we were up there. He was dead."

Specialist-4 MacCallum stated that, "When the main assault was over, I was sent down to the LZ perimeter to help with the wounded. We had taken quite a few WIAs and one KIA during the assauLieutenant Things were a little scarier down on the LZ now. The enemy had shown us that it could round the mountain and get behind us without being seen, and it would have been a good tactic to lie in wait for the medevac helicopters. Those of us that were pulling security down on the LZ were pretty much out in the open. We were nervous out there and highly alert."

When the action died down, sweeps were sent out to locate enemy dead, but most quickly came under fire. Only six NVA deaths were confirmed. According to the S-3 Daily Staff Journal report of 1/22, "Among the personal gear on the enemy bodies were found some new uniforms, small bags, letter diaries, handkerchiefs, and scarves (possibly from wives or sweethearts). For headgear most of them were wearing camouflaged hats and carrying potato masher grenades. Yesterday the enemy was carrying heavy packs but today they had no packs. Captured were one SKS rifle, two AK-47's and one RPD machine gun. All foot gear was Ho Chi Minh sandals."

"One of the KIA's did not fit the description of a Vietnamese or Montagnard. He had very yellow skin and was smooth complexioned. He was about 5'3", 120 pounds and 18-19 years old." PFC Toby Van Skike remembered seeing, "A very large NVA soldier and a white guy in weird uniforms, perhaps Chinese and Russian. After all, the enemy must have had advisors also.

MacCallum remembered, "Another platoon sent out a squad to sweep in front of their position and almost immediately they took heavy fire and had to crawl back to the perimeter. One of the men was hit by a bullet right in the forehead. The bullet went through his steel helmet liner right on one of the fastenings that hold the webbing in. The bullet was lodged in the fastening, protruding out both sides. We again came under heavy fire from the enemy."

At about 1330, Captain Burke, his wounded RTO, Specialist-4 Pete Becker, Lieutenant Snellings, and Corporal Stanek were in a small crowded hole adjacent to a large tree. During an attack, PFC Diepenbrock jumped into their little sanctuary. As Snellings maneuvered to make room, he was forced to put his right leg on Stanek's chest, just as a B-40 rocket exploded above their heads.

Stanek stated, "The force of the deadly explosion was beyond comprehension. My Lieutenant had just taken a quarter-size piece of shrapnel below his right kneecap which broke his leg. My life was spared once again."

"My Lieutenant's head slumped over and I grabbed him by the neck and lifted his head to scream at him to see if he was alive. He screamed back in pain but I could hardly hear his voice from the exploding sound of the B-40 rocket. As his blood soaked my fatigues, he was able to tell me to take charge of the artillery. Fortunately, the radio had not been damaged by the blast. Now, here I was, a young nineteen-year old kid doing one of the most important jobs in the battle. I had to take care of the D Company men with my close in artillery strikes. Lieutenant Snellings would later be evacuated and was transferred to a medical facility in Japan.

Elsewhere, during 25 April, A/1/22 moved towards the mountaintop coming in from the north. Along the way they located several fresh enemy bunkers, although their enemy contact had still been minimal.

PFC Salber stated, "I eventually did get hit about 1420 that afternoon by a B-40. The blast threw me over the sandbags and several other guys were also hit at the same time. I helped them the best I could. I was hit in the neck, arm, and jaw. My rifle sling was cut in half and the stock had little burn holes all over. My helmet was full of dents and my flak jacket was ripped up. I must have looked like hell with blood and dirt all over. This really wrecked my day, but I knew it wasn't time to quit. With good fortune, one of the life-saving medics reached me and got my bloody wounds wrapped up."

Above the battlefield and through the tree canopy, a Cessna 0-1 Bird Dog spotter airplane flew, seeking out enemy locations. When the Army troopers heard the engine of the single propeller piper cub above the raging battle, they attempted to throw smoke grenades to show the pilot where the NVA were located and where their own friendly lines had been established. Unfortunately, the enemy also had smoke grenades and they tried their best to trick the pilot by throwing out the same color of smoke. Eventually, through radio communications and the smoke grenades, the pilot was able to find his target area and fired off his smoke rockets to mark the kill zone for waiting fighter jets. The

F-4 Phantom jets came pouring in low and fast and unleashed their destructive bombs upon the hidden soldiers.

"The close in air support was like in the movies. You could hear the 'pop' on the aircraft when they dropped the bombs and hear the trees hit by the wings. Those guys were damn good pilots," recalled Van Skike.

PFC Salber was next recruited to help direct an incoming napalm airstrike. He was told to "get in tight and stay down." The jet flew in fast, loud and low and dropped its liquid fire right in front of his position near the perimeter. Someone yelled out, "How was that Roger?" He yelled backed to the command post, "That f***** bomb was too close," as the fire ball explosion sent its flames close to his face. He had helped to keep the close-in enemy from overrunning their position. He also helped to bandage up several wounded comrades.

PFC Salber reported, "Thanks to napalm, air strikes, and artillery. We had no chance without it. We dug in after dark. My squad was on the northwest side of the perimeter."

B/1/22 finally made it to the top and linked up and integrated into the D Company lines. Corporal Stanek recalled, "The men of B Company shared their C-Rations and water with us since we had had little to eat or drink for two days."

Sergeant Bliefernicht recalls: "That morning we got the word 'saddle up.' We are going to relieve D Co. We humped to D Co.'s location. When crossing D Co.'s L.Z. I saw a pile of our dead G.I.'s. Two of them I recognized, one was Fred Sena Jr., a Pia Indian from Pueblo, Colorado who I went through A.I.T. with at Fort Polk. Fred didn't have to be there, he was underweight and could not enlist so he went to work at a restaurant to gain weight. He told me he was going to get his family off the reservation by doubling his insurance or going to college. I learned in the last few years Fred refused to be Medi-evaced and let others go before him. The second person I could be mistaken about, but I believe he was the only son in his family. His Dad said my son is no better than anyone else. Both these guys I went through basic and A.I.T with."

"Whenever Stanek rested from bringing in artillery fire," recalled Specialist-4 Lynn Shadoan, "I would then take over and call in 4.2 mortars to provide protection for our hilltop positions." Stanek related, "I have had to call in hundreds of 105 rounds by this time, yet the enemy constantly moved through the jungle to locate another spot to kill from."

Casualties mounted up and by day's end, D Company still held on, but they had 29 wounded and 2 killed (Staff Sergeant Karl Lucas and Specialist-4 Fred Sena Jr.). Specialist-4 Fred Sena was a Pia Indian from Pueblo, Colorado. He had joined the Army to make his father proud that he had become a brave warrior. Sena was the best point man in D Company and was truly a courageous soldier. NVA bullets ended his life on the mountaintop that day and he would be greatly missed by his fellow warriors.

"I continued fighting for several hours until loss of blood did me in" Salber recalled. "I was cable-lifted out about 2000. The dust off had 17 bullet holes in it when it landed at the field hospital." He landed at the 71st Evacuation Hospital in Pleiku where General Westmoreland presented him with his Purple Heart.

Salber received the Army Commendation Medal, Purple Heart, and the Combat Infantryman Badge (CIB) during his first two days of combat in Vietnam. These two days had been ones that he would never forget.

The combined support received on 25 April included: six gunships, seven air strikes, 1,201 artillery rounds, and 421 mortar rounds.

Meanwhile, C/1/22 carefully advanced up the mountain from the southeast and observed lots of enemy movement all day. Eventually, they pulled back and waited while artillery and aircraft pounded the area to their front. At 1803 hours, C Company heard mortar tubes popping and shot an azimuth to it and reported the enemy is in the saddle of coordinates YA 843886.

So far, only D/1/22 had been bloodied, but that would soon change.

Lieutenant John Bobb recalls, "Our company formed a perimeter on a flat clearing prior to a steep wooded incline that would have to be negotiated successfully to assault the main portion of the hill to take the pressure off D Company.

"My platoon was assigned perimeter security, 3rd platoon flank security on the way up the hill, and 2nd platoon had point.

"In the way of explanation, when we were on company sized patrol, and then stopped in a night location to set up a perimeter, the way it worked was the platoon on point would man the perimeter in the direction of travel, as on a clock, ten to two, the next platoon would set up from two to six, and the last platoon would set up from six to ten. The platoon sergeant would then check the platoons to make sure each platoon was linked up with the other. The platoon sergeant, once this was accomplished, would then report to the Lt Platoon leaders, who would check the location for bunkers to be dug, and report to the Captain for instructions. Then all kinds of things would happen at once. The forward observers would start pre-plotting mortar and artillery fire for defense of the perimeter. Each platoon would send out a fire team sized defensive patrol to sweep the area in front of each platoon's bunker area for signs of the enemy, and then each platoon would send out small observation posts in the day, and called listening posts at night.

At our infamous Chu Moor perimeter, the first platoon was facing the hill at ten to two; the second platoon was two to six, with the famous big tree and chopper LZ in its area towards the valley in the direction of the firebase. The third platoon was six to ten, facing the unknown, enemy side, valley and mountains."

Meanwhile, the 14th Infantry's Golden Dragons units of B and C Companies discovered several bunker complexes during the day and a fifteen-foot wide camouflaged road going north to south in the area north of Chu Moor. There were indications that chain saws and building equipment had been used to construct the road.

Companies B and C 1/14 was assigned to continue to sweep the road and the adjacent high ground. A and D 1/14 were scheduled to conduct local patrols around LZ Swinger.

DAILY STAFF JOURNAL OR DUTY OFFICER'S LOG

ORGANIZATION OR INSTALLATION	S-3, 1st Bn 22nd Infantry REGULARS
LOCATION	
PERIOD COVERED	0001 24 Apr 68 - 2400 24 Apr 68
ITEM NO. / TIME	INCIDENTS, MESSAGES, ORDERS, ETC.

ITEM NO.	TIME	INCIDENTS, MESSAGES, ORDERS, ETC.
		SUMMARY: GOLDEN DRAGONS continued operations in assigned AO. Bravo and Charlie Co's found several bunker complexes with bunkers built to withstand 750 lb. bombs. They also found a camouflaged road approximately 15 feet wide going N&S. Indications are that chain saws and building equipment were used to complete the road. PLANS: B&C Co's are to continue to sweep the road and adjoining high ground. A&D Co's to conduct local patrols.

4TH INFANTRY DIVISION OPERATIONS SUMMARY
21 APR 68—30 APR 68

25 April. At YA944891 Company D, 1st Battalion, 22nd Infantry killed two NVA snipers in trees and captured their AK-47's. An hour later the company received mortar fire and shortly thereafter made heavy contact with an estimated reinforced NVA company. The contact continued through the morning, and the company received two B-40 rocket attacks in the afternoon. Results of the contact were two US KIA, 29 US WIA, and six NVA KIA. At YB934832 the 206th MSAF company at DAK PEK made heavy contact with an unknown size force. The company became separated into three groups during the fighting, which continued for more than four hours. Results of the contact were 10 MSWF WIA, 14 MSF MIA. LRP 3C found one NVA body at ZA147942. At ZA060925 a 5 ton truck from the 5th Battalion, 22nd Artillery struck a mine, resulting in two US WIA, one US MIA. At AP822981 the 3rd Battalion 45th ARVN engaged an unknown size force, resulting in five ARVN KIA, 11 ARVN WIA, one ARVN MIA and 23 NVA KIA.

CHAPTER 7
Day 3: Friday 26 April 1968

"I then heard shooting, screaming, and then support gunship helicopters making strafing runs."

DAILY STAFF JOURNAL OR DUTY OFFICER'S LOG

ORGANIZATION OR INSTALLATION		S-3, 1st Bn 22nd Infantry REGULARS
LOCATION		FSB YA868888
PERIOD COVERED		0001 26 Apr 68 - 2400 26 Apr 68
Item No.	Time	Incidents, Messages, Orders, etc.
10	0718	B Co received B-40 rockets from NW.
12	0725	B Co 1 hurt at 0718 by B-40 shrapnel in head. Required dustoff.
17	0826	D Co receiving incoming rockets & ground fire from NW.
18	0827	About platoon size enemy, B-40 hit bunker have some injured.
19	0828	D Co in heavy contact, request gunships.
26	0845	D Co CO says that it is quiet at this time.
28	0905	D Co has 3 more WIA's.
29	0908	B Co had 1 man dusted off so far.
31	0914	D Co in heavy contact, ground fire from NE.
32	0916	Enemy 100m out of perimeter; A/1/12 SP.
34	0925	C Co has movement to the front.

38	0940	Airstrike completed; 1st for the day.
41	1001	D Co 1 of their WIA's is now a KIA, named Carol.
42	1010	2nd airstrike completed at this time.
44	1020	C Co reports has found old MY grave, old punji pits and some recent signs.
46	1041	C Co has movement again to front, still moving.
48	1053	Dustoff 34 completed dustoff of 4 NVA.
51	1057	Co B 3rd platoon believes he has incoming rounds.
53	1102	B Co 1 WIA—was wounded while on way to LZ.
56	1116	B Co has 3 WIA, D Co has 7 WIA & 1 KIA.
59	1135	3rd airstrike completed at this time.
60	1140	D Co has 1 more WIA, total of 8 WIA & 1 KIA.
61	1144	B Co 3rd platoon has incoming, has wounded.
62	1150	B Co 3rd platoon has 4 seriously wounded men.
63	1151	B Co starting sweep now.
64	1155	D Co receiving small arms fire from N & NW.
65	1156	C Co has heavy contact from small arms fire from north.
66	1158	C Co has 1 WIA.
68	1200	D Co has 60mm incoming mortars and B Co 3rd platoon also receiving incoming mortars.
69	1204	Mortars believed to be coming from SW.
70	1209	Bn 6 informed Bde 6 that he believes enemy force is battalion size.
71	1212	B Co has 3 WIA and 1 WIA of D Co at PZ.
72	1214	Cougar 13 spotted bunkers and a practical shack to the SE of the element.
73	1216	C Co now has 1 WIA & 1 KIA.
75	1220	A Co location YA845895, heard mortar pop 300' azimuth from there.
76	1222	Cougar 13 heard mortars pop in area—will fire in area.
77	1239	C Co CO believes 1 of guns received ground fire. Has been hit be ground fire, 1 pilot shot up bad.
78	1245	Dustoff will pick up wounded pilot in 7 minutes.
80	1248	A Co CO informed us he heard tubes pop from same azimuth.
84	1302	Dustoff of pilot completed at this time.

86	1312	Cougar 13 was 20' above tree level and received AK-47 rounds.
87	1313	D Co LZ has 8 WIA's to be dusted off, 4 are urgent. C Co has 2 WIA's to be dusted off & 1 KIA.
88	1316	Cougar 14 took 1 hit in his bubble and the pilot got hit by the round.
92	1400	D Co has heavy contact at this time.
95	1404	Gladiator 059 was fired at while leaving C Co LZ.
96	1405	B Co shot a sniper out of a tree.
101	1422	D Co CO says he has sniper fire all around him.
102	1428	B Co & D Co combined, 11 WIA's to be dusted off, and 2 are litter cases.
104	1437	Bde 44 says airstrikes for 1515/1545/1605.
113	1603	Dustoff of all WIA's is completed at this time.
114	1605	B Co receiving incoming mortars and ground fire at this time.
115	1607	C Co had movement to their front and reconned by fire.
119	1627	Airstrike completed at this time.
121	1640	C Co reached top of hill and ready to go over.
122	1647	B Co received 1 B-40 & small arms at this time.
125	1708	C Co in contact at this time YA842888.
126	1715	C Co has a couple WIA's. Don't know how many yet. Believes enemy is dug in and doesn't think airstrikes got to them.
128	1730	C Co has 2 WIA.
129	1743	C Co has 4 WIA's to be dusted off and 1 KIA to be taken out.
133	1758	B Co—8 WIA's, 0 KIA. C Co—4 WIA's, 1 KIA. D Co—9 WIA's, 1 KIA.
138	1827	Gladiator 5 received ground fire N of DZ of C Co.
145	1905	B Co had 15 WIA's—9 dusted off and 2 to be dusted off tomorrow. 1 of them belongs to A Co.
151	2200	From Bde CO: Indications are that the 66th NVA Regiment is prepared for combat in the 1/22 AO. The location of the 66th Regt is YA8289. Their combat operations will probably take place in the area of contact today.

152	2205	From Bde: regarding secondary Arc light targets HA 416 grid: YA710830, YA7208930, YA710850, and YA720850—no bombs before 270210 or after 270320. HA grid: YA745855, YA739863, YA763875, YA754880— no bombs before 270650 or after 270800.
155	2231	Request to Bde: (1) request airstrikes on a continued basis. We have known enemy targets and they will be needed to support maneuvers. (2) request 7/17 Cav conduct operations in 1/22 AO tomorrow. (3) request 2 quad 50's to enhance FSB perimeter defense. (4) consideration given to planning LRRP vicinity YA8488, 888, 8487. Primary area of entry is the valley running N & S. (5) Arc light at YA881887, 811907, 801907.
161		Summary: 1st Battalion 22nd Infantry with A/1/12 Infantry Opcon continued recon and force operations in assigned AO with heavy contact. A, B, C & D Co's 1/22 heavy contact with Battalion size enemy force. Received B-40 rockets and mortar fire and small arms. Airstrikes and Arty and Gunships were employed.

As dawn struck the Chu Moor mountaintop on 26 April, both B and D Companies of 1/22 began to take incoming rockets, mortar rounds, and small arms fire from all around their perimeter.

Sergeant Arlen Bliefernicht (B/1/22) recalls, "When we took over D Co.'s position, all was quiet for a short time. Than a chopper approached our L.Z. and the snipers opened up on the chopper. The chopper was driven off. Sergeant Cowley, our platoon sergeant, said, 'I have had enough of those snipers'. He grabbed an M-60 Machine gun and a couple belts of ammo. He took care of two or more of the snipers, he shot one up so bad, that there were pieces of the sniper that fell from the tree he was chained to.

"Meanwhile things were heating up out to our front. We had mad minutes firing everything we had at their bunker line and of course they would return the favor with a lot of B-40's.

"When we arrived, there was only about ten yards cleared for fields of fire. When I was Medi-evaced a few days later, the fields of fire were about 70 to 80 yards out, all cut down by bullets from both sides. One person remarked that someday a future lumberjack was going to be mad at us when his chainsaw hit all those bullets we put into the trees. Whenever a small tree or bush would go down, someone would yell timber or another one bites the dust. A little bit of humor kept our morale up.

"When evening was arriving, we noticed the NVA were cooking their supper, from smoke arising from their positions. We looked at each other, let's eat our chow. In the following days, the same thing happened, in the morning, they had breakfast; we had breakfast, followed by many mad minutes. The NVA never attacked us at night. I guess both sides wanted a good night's sleep and a full stomach before killing each other."

Casualties began piling up and medevac's were brought in as helicopter availability and conditions allowed.

"A B-40 rocket blew away my machine gun squad," recalled Specialist-4 Jim Dillard (D/1/22). "I assisted the medic as we went over to a sergeant, PFC Tom Carroll, and Specialist-4 Rainer Guensch. Tom was less coherent than the remainder of the team. He had a leg wound and deep shrapnel wounds all over his back. He said he couldn't breathe and it wasn't too long before he died."

Rainer Guensch had been leaning against a fallen tree, next to his foxhole, with a small squad of men, when an incoming B-40 rocket hit near their position. As the dust cleared, it was evident by the groans and screams that most of the men had been wounded, including Guensch who had a two-foot long gash on both of his legs. Pissed off, he grabbed one of the enemy weapons that he had retrieved from the day before and emptied an AK-47 magazine towards the NVA positions. Soon, medics made it to their positions and began patching the team up and moved them down to the helicopter zone. From there, Rainer remembered that Jim Dillard and another man placed him in a basket and being lifted up towards an awaiting helicopter. The basket twirled around and around as it maneuvered through the trees and branches. "I knew that my injuries were going to keep me in hospitals for a while and that I would not get the 'early out' that I had been expecting!" Dillard remained at the LZ and continued to help with the evacuation of the wounded throughout the day.

Company C 1/22 continued its uphill march towards the mountaintop, finding an old Montagnard grave, old punji pits, and some signs of recent activity. Company C continued to have movement to their front and finally made heavy enemy contact just before noon. Within 20 minutes, Company C suffered one WIA and one KIA.

Lieutenant Bobb continues, "This is where some formation and Captain Konnerman's tactics come in. There were different formations depending on terrain and threats, single column, column of two's, and what a favorite the wedge formation became. This provided better flank security, with a platoon on each side of the point platoon. If the point platoon element was hit, they could deploy for covering fire, and would then have a platoon on each side deploy for more cover fire. Theoretically instead of having three soldiers open up to suppress enemy fire, we would immediately have nine soldiers providing suppressing fire, so the point fire team could retreat into the safety of the larger unit without getting decisively engaged. The military definition of decisively engaged is being in a position where you can no longer control your actions in the battle. Meaning you can't execute a retrograde movement, frontal assault, or a flanking maneuver. You are basically in a killing zone of an enemy and have lost the ability to influence the outcome of the battle in your favor. I went into all of this because it is important later.

"I then heard shooting, screaming, and then support gunship helicopters making strafing runs. I observed Lieutenant Westbrook and three of his soldiers carrying a dead soldier down the hill. I had never seen a dead person before.

"Picture our perimeter. We had a one chopper LZ or PZ, the chopper had to semi-land forward, then back out, and then turn around. This was our life line for evacuating casualties and getting ammo and water, etc. What would have happened if our entire company would have just been put on line and did a frontal assault on that hill, being outnumbered, without protecting a perimeter or that little LZ.

"The hope is that the enemy would leave between prepping the hill with all kinds of fire support, and our subsequent assaults, because they usually did some hit and run stuff, and would take off before the fire support came in. This time they didn't leave for whatever reason. Of course we couldn't read their minds. I think some higher ups wanted us to be more aggressive, but you should never conduct a frontal assault on an entrenched enemy unless you have a minimum of a three to one advantage. I guess the higher ups can figure how an enemy is armed, and substitute our numbers because of the available fire support; however that is above my pay grade. There was nothing stopping higher ups from throwing another company or two our way.

"They had D, A, and B companies going up one direction because D Co had been in the shit the longest, and we were on an opposing ridge by ourselves. I personally don't remember any of the big picture info. I did have a D Co. soldier tell me they could hear our firefight, and that they would all be dead if it wasn't for us. I have no knowledge of anything."

Specialist 4—Sam Drake (C/1/22) remembers, "We moved up the mountain from the valley floor, C Company was informed that D Company was in contact with a sizeable NVA enemy force. We were instructed to join with D Company if at all possible. The mountain was extremely heavy with the jungle canopy. About three-quarters of the way up the mountain side; we were spread out in two to three files out making way through the dense jungle when we were fired upon by the NVA. I was far to the left when we were hit; however, most of the engagement was to the right of me. I was nearest to a young soldier whose name was Truman Lovins. He was hit in his right arm, medic Shyab moved up under fire to give aid to him; he helped Lovins off the hill to the LZ for evacuation.

"We moved back down the hill about 75 to 100 yards to a location that was nearly flat with a very large tree in the middle of our perimeter. C Company called in an air strike in the afternoon. We then started digging bunkers and clearing a landing zone. As I recall, that night there was no known enemy activity near our bunkers. We dug our bunker at the base of the large tree."

Company B 1/22 received B-40 rocket fire and mortars up to 1200 and had four seriously WIA.

Staff Sergeant Ned Williams, new in country and just assigned to Company B 1/22 flew into LZ C-Rations. Upon arrival, the sergeant asked where his new company was and he was told that it was where all of the smoke, air strikes, and fighting was going on, on the mountain just across the valley. "It wasn't long before I was on a slick (helicopter) with a few other guys and supplies headed to the battle site. The chopper couldn't land, so it hovered about seven to eight feet above the ground and we jumped off.

"The supplies and body bags were thrown out as fast as possible. Jim Dillard and several other soldiers helped me put the KIA's in the black rubber bags and handed them up to the nervous helicopter crew. The door gunners were firing their M-60 machine guns randomly into the jungle to keep the enemy's heads down. All I could think was; what in the world had I gotten myself into."

D Company had finally been relieved and left the mountaintop at about 1500 as A/1/22 approached the upper perimeter to replace the seriously depleted company. "I remember that after we left the perimeter and had gone a couple hundred meters or so, B Company all of a sudden opened up with everything they had. I felt sorry for them, but my predominant feeling was relief to be getting off that mountain," recalled MacCallum.

MacCallum continued, "I distinctly remember counting the number of us who walked down the mountain. There were 27! We arrived at C-Rations at about 1800." DeVito added, "We broke through the NVA lines and cleared a path down the hill. We were all shocked and ran, fell, and jumped down into the valley and up the side of another hill (LZ C-Rations)."

At 1607, C Company had movement to their front and reconned by fire.

An airstrike was completed at 1627.

By 1640, C Company reached the first and lower crest of the mountain and was ready to go over. At 1708, C company was once again in contact at coordinates YA842888 and received two more WIA. C Company reported the enemy is dug in and doesn't think the airstrikes got to them. It had sustained a total of four WIA's and one KIA during the day's fighting.

Sergeant Arlen Bliefernicht B/1/22 remembers, "Later that day B Company was joined by A Company. A Company right away did an on line charge, led by a man carrying an M-60. As he rounded a tree he was hit in the belt buckle by a B-40, he was cut in two. We from B Company had yelled to A Company people not to go out there. We watched unbelieving as A Company people tried to recover his body, each time more people getting hurt. I always said, 'When you're dead Fred, you're dead, don't come for me.'

"I don't remember exactly when C Company came across the valley and tried to break through to B Company. B Company tried to break out to meet C Company, but we were met by very heavy firing from the NVA. We could hear you yelling and firing, but couldn't get to you."

With the arrival of A Company, there were now three 1/22 companies (A, B and C) on or near the top of the mountain. The friendly casualty totals for 26 April were 29 WIAs and 2 KIAs, with each company taking on some of the burden. Unfortunately, these casualty figures would greatly increase the following day. Once again, enemy losses could not be determined, but it was obvious that they, too, suffered greatly, especially from the air and artillery support that was provided throughout the day and evening. The support for 26 April consisted of six gunships, nine air strikes, 2,654 artillery rounds, and 229 mortar rounds.

Sergeant Arlen Bliefernicht B/1/22 recalls, "We replaced D Company after they got hit hard. It was on the 2nd day, and nobody had been able to go to the bathroom. It was about dusk and my opposite squad leader, Sergeant Clayton, said, 'I can't hold it any longer.' So he took a piece of plastic and spread it out in the bunker and did his thing. He then wrapped it up in a nice package. He then tossed it out towards where the NVA was dug in and said, *'Eat shit, Charlie.'*

"During the night we heard a 'thunk' down the line by our M-60 emplacement. We counted off the seconds and it was declared a dud. Someone said, 'Don't cook your breakfast' on top of it.

"The next morning the M-60 squad leader came to our position and approached Sergeant Clayton. He dropped his plastic bag of shit in his lap and said don't ever tell us your shit don't stink, because Charlie sure does. Sergeant Clayton replied what should I do with this now and prepared to throw the bag of shit back at Charlie's position. Couple of the guys grabbed him and said don't do that, next time they may attach a grenade to it.

"During the time I was at Chu Moor, artillery was causing more problems than the NVA. We had two or three firebases firing in support of us. One of the firebases kept on shelling our L.Z., hurting the wounded and the security men protecting them. It got so bad nobody wanted to guard the L.Z.

"So, Seeber Kelly volunteered to pull security on the L.Z. for our wounded. Kelly was a smiley, outgoing trooper. He usually was singing things like; I know we can do it. Kelly went down to the L.Z. Shortly after Kelly arrived at the L.Z. another artillery barrage hit the L.Z. Kelly was badly wounded and lost his leg. He was Med-evaced. He sent back a couple of upbeat messages. He died on his way to Japan."

The 1/14 continued to assist 1/22 by conducting search operations along the wide road that it had discovered the previous day in the northeast. A vacant enemy bivouac area that had once held approximately 200 people was found by A/1/14, although it appeared that the bunkers had not been used for several days. Next day plans were for B and C Companies to move south in support of 1/22 while A Company returned to LZ Swinger for security.

DAILY STAFF JOURNAL OR DUTY OFFICER'S LOG

ORGANIZATION OR INSTALLATION	S-3, 1st Bn 22nd Infantry REGULARS
LOCATION	
PERIOD COVERED	0001 26 Apr 68 - 2400 26 Apr 68

Item No.	Time	Incidents, Messages, Orders, etc.
		SUMMARY: GOLDEN DRAGONS continued operations in assigned AO. Charlie Co moving along the road found an old bivouac area for approx. 200 people. Had not been used recently. PLANS: Bravo and Charlie Co's to conduct CA to the south in support of 1/22 Inf. Alpha Co to move to LZ Swinger for security mission.

4TH INFANTRY DIVISION
OPERATIONS SUMMARY
21 APR 68—30 APR 68

26 April. At YA198695 a UH-1H from Troop A, 7th Squadron, 17th Cavalry received an air burst, slightly wounding three US. The aircraft landed on Highway 14S to check the damage, but had to lift off and return to KONTUM when it received fire from both sides of the road. At YA937935 Company C, 2d Battalion, 35th Infantry found three NVA bodies killed by artillery. At YA844891 Companies B and D, 1st Battalion, 22d Infantry made contact with an estimated NVA company, receiving SA, AW, B-40 and 60mm mortar fire. Two hours later Company C also made contact with an estimated company at YA843886. By 1250 hours all three companies were heavily engaged. At 1255 hours a gunship from the 52d Aviation Group received heavy ground fire, seriously wounding the pilot. At 1600 hours Company A conducted a relief in place of Company D. At 1800 hours Companies A and B reported the enemy was withdrawing. At 1845 hours the enemy broke contact with Company C. Results of the contact were two US KIA, 28 US WIA and four NVA known killed. At ZA162167 a patrol from Troop A, 1st Squad-

ron, 10th Cavalry received SA fire from an estimated three individuals, resulting in one US WIA.

CHAPTER 8
Day 4: Saturday, 27 April 1968

"He and I looked at each other and I said, let's do a John Wayne."

DAILY STAFF JOURNAL OR DUTY OFFICER'S LOG

ORGANIZATION OR INSTALLATION		S-3, 1st Bn 22nd Infantry REGULARS
LOCATION		FSB YA868888
PERIOD COVERED		0001 27 Apr 68 - 2400 27 Apr 68
Item No.	Time	Incidents, Messages, Orders, etc.
01		B Co reports receiving incoming rockets. Request Spooky. No casualties.
06	0215	B Co reports they had more movement to the west. They reconned by fire and received some incoming. Advised that Spooky 22 enroute.
09	0352	Spooky 22 has finished working for B Co and has departed the area.
10	0400	B Co reports more movement, request Arty be cranked up.
11	0418	B Co receiving B-40 and small arms at this time.
12	0420	B Co reports incoming stopped at this time.
13	0421	B Co reports they're receiving incoming mortars.
15	0430	B Co reports incoming mortars stopped.
22	0505	Spooky 23 checked in at this time.
23	0606	Spooky departed the area at this time.
25		B Co receiving incoming but Arty is effective.

26	0645	B Co reports all is quiet at this time.
27	0650	B Co just received 1 incoming B-40 rocket.
30	0715	A Co reported receiving B-40's on azimuth of 40'.
36	0808	Gambler 9 checked in at this time.
38	0820	A Co has 2 WIA's—were wounded at 0715 hrs.
44	0857	Tornado 6 and guns will be working to the west.
49	0920	Tornado will be working to the south and will be on this push.
50	0934	Headhunter will be working in our AO, in the contact area & west to the 76 line; north to the 92 line; south to the 83 line.
54	0950	Airstrike completed at this time. Enemy still up on terrain feature.
57	1002	C Co has 4 WIA's at this time.
60	1012	C Co still in contact at this time. Has snipers and some enemy in bunkers on top of terrain feature.
62	1035	Gambler 6 checked in to work for us.
63	1036	C Co receiving mortar fire, mortars coming from W of C Co.
64	1040	C Co reports mortars are coming from azimuth 310' at 600m distance.
65	1044	B Co got 2 more NVA KIA's / total of 4 now.
67	1055	Dustoff completed with 3 WIA's & 1 KIA.
68	1100	Gambler 9 & guns checked in at this time.
69	1103	C Co has a WIA that cannot be pulled back to safety. C Co CO believes he may be a KIA—MIA.
71	1111	White 25 going in for C Co to expend.
72	1114	White 25 received ground fire on his pass.
75	1139	C Co had incoming mortars (3/4 rounds).
77	1218	B Co received another rocket, 320' azimuth—1 WIA.
78	1228	B Co has another WIA from rockets.
79	1235	Airstrike completed at this time.
80	1256	C Co still progressing up to objective. Are firing at what appears to be a sniper in a tree.
82	1312	Tornado White 25 checked in for support.
83	1314	C Co receiving mortar rounds following him up trail. C Co has heavy movement to right flank.

84	1316	Gambler 9 checked in for support.
87	1341	Gambler 11 received some ground fire on last pass.
88	1405	C Co receiving incoming mortars (60mm).
89	1413	B Co reconned by fire & received AK-47 return fire.
90	1415	C Co receiving AK-47 fire and guns receiving ground fire.
91	1415	B Co has intense fire also, has confirmed NVA WIA.
92	1419	C Co received 60mm mortars at N/L on a 310' AZ & 600m from PZ.
95	1500	C Co has 5 WIA, 1 KIA, 2 MIA.
96	1501	B Co receiving small arms fire from SE.
97	1508	A Co 3rd platoon got 1 NVA KIA sniper confirmed.
98	1526	A Co has 2 WIA & 1 KIA results of contact.
99	1530	C Co N/L has incoming at this time.
103	1536	C Co N/L mortar from AZ 310', 3 seconds from tube until it hits.
104	1537	A Co 1st platoon got a sniper, NVA KIA.
107	1551	B Co receiving B-40 rockets at this time.
109	1602	Our S-1 informed us that B Co has 3 WIA's & A Co has 3 WIA's.
110	1604	Dustoff 37 completed dustoff of 3 NVA, 1 B Co, 2 A Co.
111	1609	A Co has 2 more WIA's.
112	1610	C Co has 1 WIA.
113	1617	C Co reports receiving 4 mortar rounds at N/L.
114	1631	C Co 1/14 reached top of nob, still checking bunkers, will be up on push in 15 minutes.
115	1641	C Co has reached top of objective.
117	1646	C Co has contact on top of objective at this time.
118	1650	C Co receiving mortars on top of hill at this time.
120	1654	B Co receiving B-40 & has position pinpointed & is calling Redleg at this time.
122	1656	C Co on top of objective has enemy on northern side of hill and on his east & west flanks.
123	1658	A Co receiving firing in contact area. 3rd platoon confirmed NVA KIA.
125	1709	Tornado White 25 departed location. Tornado White 23 on station at this time.
126	1710	B Co received 1 B-40 rocket at this time.

128	1721	A Co has 4 WIA due to the B-40 rockets received at 1710.
129	1727	A Co has 7 WIA's; B Co has 3 WIA's & 1 KIA.
131	1733	A Co received 1 B-40 rocket.
136	1800	Cougar checked in at this time.
139	1813	C Co in heavy contact at this time on top of their objective, have pulled back a little.
140	1817	Dustoff 38 completed dustoff of 6 NVA.
143	1841	Dustoff 37 received ground fire at this time, negative hits.
145	1910	C Co got a secondary explosion on the mortar site the 4.2's were firing on.
146	1916	C Co reports bunker location YA843885—there are 2 machine gun bunkers & 1 bunker with no MG. On crest of hill, bunkers on top of hill continue & SE corner of next contour, 1 bunker top of hill, 1 middle, 1 SE flank—before Dustoff.
148	1941	C Co reports that they had 2 more secondary explosions on mortar site.
149	1945	NVA KIA killed per Co: A—2 NVA; B—2 NVA; C—1 NVA.
152	2030	C Co CO indicates he has movement to his S.
153	2050	Casualties for today: A Co—19 WIA's, 3 KIA's; B Co—18 WIA's, 1 KIA; C Co 13 WIA's, 1 KIA, 2 MIA.
154	2053	Enemy casualties—5 KIA & 4 KBA.
160		Summary: 1st Battalion 22nd Infantry continued recon and force operations in assigned AO, with contact with enemy battalion. A, B & C Co's had heavy contact, received B-40 rockets, mortars & small arms while attempting to sweep contact area. 4.2 had 1 secondary explosion. Artillery, airstrikes, gunships & Spooky employed. Results of enemy casualties are unknown. Friendly casualties—53 WIA's, 4 KIA's, 2 MIA's.

In the early morning dark hours of 27 April, A and B, 1/22 started hearing movement outside their lines and began to take heavy rocket, mortar, and small arms fire. Besides requesting more artillery, they also received the welcomed assistance from AC-47D's Spooky 22 and 23 gunships.

Sergeant Bliefernicht (B/1/22), "April 27 started out like any other day on Chu Moor, we had breakfast, and they had breakfast. Then we started to try to kill each other, with a couple mad minutes. It was between one of these mad minutes, I was in the bunker line getting ready for the next mad minute when I was slammed back on my back, it was like a giant fist had hit me. I screamed for a medic.

"The medic arrived and tore open my shirt, I couldn't look, and I felt my shoulder was torn off."

"The medic then said to me, 'Sergeant, bandage it yourself, you big baby'. I then looked at my shoulder. There was just a small trickle of blood, but it had hit my collar bone. Boy did it ever hurt. A B-40 had hit the tree in front of my bunker and had left a hole in the tree the size of a golf ball. By the next morning my shoulder was black and blue from my elbow to my neck, and I couldn't raise my arm. I showed the medic my shoulder and he apologized for his remark. I had become a walking wounded, and went back on line with the rest of my squad.

"We realized the NVA had one very good sniper above us in the trees. He would only shoot when we were doing our mad minutes and was a crack shot. PFC Robert Williams was in my squad and was only in country two weeks. Williams was next to me in our bunker, when he took a round in the head. Medic was called and declared him dead. The Medic looked at me and said, what's that gray stuff on your shirt; it was parts of his brain. I brushed my shirt off and continued to fight.

"During that day the NVA sniper took out a few other men. We fired thousands of rounds trying to get that sniper. Later that day, the air force was bombing and strafing in front of our location. One pilot held his strafing run about two seconds too long, and killed that sniper. The pilot radioed that he hoped he didn't hurt one of us. No, we said, but you got that sniper we have been hunting for the last couple of days. The sniper was chained in a tree a few yards out, he was really good. I had to admire his skill and courage.

"That morning our C.O. ordered our Lieutenant to do an online charge towards the NVA's bunkers. Our Lieutenant was very new in-country, he opened his, 'how to do an on line charge book'. The Lieutenant was standing up in front of our bunker and telling each one of us which spot we should do on an on line charge. All of us were yelling, this was not the way to do this. The NVA must have been surprised too because they didn't open fire for a few minutes, lucky for us. We all dived back into our bunkers. The Lieutenant agreed this wasn't cool and threw away his 'how to do an online charge book'.

"I believe shortly after this Bruce Thibodeau and I looked at each other and I said, let's do a 'John Wayne.' People do crazy and stupid things in combat and this is one of them. Bruce said cover me and ran to the first tree, I did so, and followed Bruce. I covered Bruce as he and I ran from tree to tree, covering each other as we went. Everybody in the company was yelling for us to get our butts back to our bunkers. As both of us got behind a big tree right before the NVA bunker line, bullets were hitting the other side of the tree. It sounded like a bunch of woodpeckers on the other side of that tree. I looked at Bruce and said let's do a 'John Wayne' in reverse. We leaped frogged back to our bunkers. What neither of us knew was a B-40 hit the tree as soon as the last one of us left… three trees, three times."

Lieutenant John Bobb remembers, "On April 27, it was my turn to take point. The usual mortars, artillery, gunships, and F-4 Phantoms pounded the top of the hill. My most aggressive NCO, Hollis "Pappy" Buck, went up the hill fast so the enemy would not have time to get back in their fighting positions, and we started getting on line for the assauLieutenant Fast didn't work.

"Pappy and Howard Hominick were killed immediately by machinegun fire. Pappy had been in Korea and stepped on a landmine. Pappy had a scar up the entire length of his spine. He was told he would not walk again. I guess he told the doctor he was in the infantry. Pappy was loved by all and was 36 or 38 years old. The firefight was ferocious. Robert Roark, now thrust into a more senior leadership position, was killing NVA in trees, and firing towards enemy machinegun fire in order to cover our machine gunner, Fat Terry, so he could seek cover behind a downed log. The NVA then turned the log into sawdust. I knew Roark was in charge of the right side, and I got up and checked the left side. I found my other machine gunner, Joe Cox. I needed to make sure we didn't start shooting each other until we located the remaining members of the lead fire team that had survived the initial ambush. Cox was being cool and maintaining fire discipline.

"I returned to my original position and remember having to cover each other, just so we could shit. One time it was our 1st platoon's turn to man the perimeter. I remember sitting in a little shade, leaning on my rucksack, and I took out a paperback book, and held it in my hand after I had checked the perimeter. I knew that when the other two platoons went up the hill I would hear enemy soldiers and friendly soldiers screaming when they got hit. I thought if I held the book in my hand I could block out the screaming. Not very rational, huh?

"Our fire support and F-4s starting coming in and I was sitting in the sun. I moved my rucksack over about two feet to sit in in the shade when I heard all of our bombs dropping, and I heard a thud, and I saw a huge piece of bomb shrapnel stick in the ground where I had been sitting a couple of seconds earlier. It would have gone right through my back.

"I remember the NVA rolling hand grenades down the hill, and I would bury my face in the ground, and put my fingers in the inside of my helmet liner, to hold my helmet on to protect myself from shrapnel. As soon as we did that and had stopped shooting in the trees, then I would see bullets landing all around me, and would have to roll over and shoot in the trees again.

"I remember walking from the company CP past the 2nd platoon area, a staff sergeant was sitting in his bunker, facing in towards the perimeter, and I kept seeing this black skinny arm come out of the bunker, and hand grenades flying in the air, and rolling downhill past his bunker and exploding. I asked what are you doing, Sarge? I looked closer, and he was sitting on a case of grenades, with another one between his legs, and he was opening the canisters, pulling the pins, and throwing the grenades backwards over his head. He said, 'Lieutenant, if they come in this perimeter, they aren't coming through this way.'

"At some point, they sent us a 90 mm recoilless rifle and some flechette rounds. If memory serves me right, there were either 600 or

900 little darts in each one. I put Louis Kimbrell on the point bunker with the recoilless, and acted as his assistant gunner. I made sure the back blast area was clear, loaded him up, and we fired three rounds up hill at different angles. I couldn't tell you if it had any effect on anything or not.

"Another young soldier was now thrust into a senior leadership position in the platoon. He was Lynn "Kelly" Berliner. He was a month younger than I was. He was a tough as nails kid from Chicago. He had already been through Airborne Ranger training, was with the 82nd Airborne, and had been in a conflict in the Dominican Republic already.

"The remainder of the fire team up front had a difficult decision to make, and they made the right one. We couldn't see anything ahead of us and couldn't hear much due to the fire fight. They couldn't even reach up and grab an extremity of the two dead soldiers due to the volume of fire. They had to let us know what had happened, and could not stay in our fields of fire. We would have to recover the bodies after regrouping. I had two very experienced RTO's while there, Elmer Hale and Sam "Rufe" Johnson. They had come to Vietnam at the same time and took care of me.

"I got the bright idea to get the LAW (light Anti-tank Weapon), a disposable rocket launcher one of my men carried. I figured I was the best trained on the LAW and I retrieved it, and was going to simply crawl forward with it, open it up according to the instructions, shoot it at the enemy machinegun bunker, recover my soldiers and sit on top of the conquered hill, and smoke my first cigarette like a grizzled old veteran.

"RTO Sam Johnson, and soldier Louis Kimbrell, from the Kansas City area, made it to my location. These two soldiers were only surpassed in their abilities as soldiers by the type of persons they were.

Sam was from the Southeast, tall and lean and strong. I told him what I was going to do, and he grabbed me and told me no. He was experienced, he was in the front, and he knew there was nothing to shoot at. I didn't but he did. We had evidently come into their house, they lived there, they were prepared, and they weren't leaving. They were camouflaged. They were in trees, under trees, and all over. Sam saved my stupid life.

"Further decisions were taken from me, as I was ordered to pop smoke for gunships. I remember screaming into the radio to not run the gunships over my men.

"Louis Kimbrell knew where my two dead soldiers were more precisely than anyone. I won't say much about Louis except for this; above all human beings born, you would want Louis to be your father, brother, son, son-in-law, friend, and to watch your back when things were bad. There may have been other human beings born as good as Louis, but I doubt it.

"For the rest of the day, we continued to assault the hill, taking turn on point. Louis Kimbrell went up first, no matter which platoon was on point. I don't know how he survived.

"The number of times and circumstances each time is a blur. I do remember reaching the same approximate point each time. I remember one occasion seeing my men on their backs, shooting into the trees. I remember seeing bullets hitting the ground all around us, and when I looked up the hill, enemy hand grenades were rolling down the hill at us. I remember my CAR-15 jamming and putting my hands in the web lining of my steel pot liner to hold my helmet on and putting my face in the dirt, hoping the grenade shrapnel would hit my helmet or go over me. Lieutenant Westbrook, had an enemy grenade land on the back of one of his soldiers, roll right into his (Lieutenant) chest, and not detonate.

"We were taught in OCS that in order to successfully launch a frontal attack on a well-entrenched enemy, you needed at least a three to one advantage in personnel to sustain fire superiority and minimize casualties.

"The F-4 Phantom Bombing runs were so close that we were instructed to lie on the ground with our elbows between the ground and our lungs, put our fingers in our ears, and open our mouths as wide as we could when the bombs hit. This was to protect our lungs and ear drums. I will admit, I had never seen fully grown mature trees, roots and all, fly over my head before this."

C/1/22 made a serious contact within an enemy bunkered area of unknown size. "On the morning of 27 April we sent two platoons up to recon the hill," recalled senior medic, Specialist-4 Charles Shyab. "We went as far as the bunker line and we withdrew under fire with four KIAs and six plus WIAs. We probed the hill at least two more times on the 27th with more casualties. The enemy was well dug in and had their bunkers under tree roots. Each time we went up they would try a different tactic like mortars, snipers, flanking maneuvers, and grenade throwing. During one of our assaults, I was called forward to treat a head/neck wound, but I could only try to find the bleeder to no avail, and just comforted him as he passed."

Sam Drake remembers, "About 0900 all was quiet and the sun was shining. In the AM we were ordered to move back up the hill. We slowly started up. My thought as we were forming to go would be I will be fighting for my life and it could be my last day. There were about three to four small trails that you could take after passing the last bunker in our perimeter and there was one young soldier that was in the front of me. We moved up the hill; however I don't remember the young soldier's name because he was new to the company. We continued moving up the hill when we came across a fallen tree, the soldier crossed over the tree as he was about three to four yards in the front of me when the NVA opened up on us from their bunkers and he was hit in the neck. They then directed their fire towards Sergeant Cappaletti and me while we were behind the fallen tree.

"Someone called for the medic and within seconds Charlie Shyab was by my side and told me to cover him. Charlie moved up the hill to help the young soldier that was hit by the NVA. The NVA must have moved back down in the tunnel or bunker because we didn't receive any more automatic rifle fire. Charlie exposed himself to work on the young man's neck wound but wasn't able to save him. Charlie pulled him back to our position as we moved back down the hill by rolling, crawling to get to the new location where most of C Company troops were behind us. We then called in an air strike on the NVA. The air strike was a helicopter gunship that fired rockets. We threw a smoke grenade where we thought the NVA bunkers were. The helicopter gunship was coming straight at us, where the smoke grenade marked the spot. We then moved back down the hill to our bunker in the late afternoon. The enemy was only 100 yards up the hill from us. In the night the NVA moved down closer to us into some of the trees."

C Company would end up making three assaults during the day, each time falling back to allow artillery and air strikes to hit the area. During one of these attacks, Specialist-4 Elmer Hale vividly recalled, "Staff Sergeant 'Pappy' Buck was on my right and we were going up this hill and he turned around to bring the guys up behind him to get on line. I turned to look at him and he was gone in an instant, ripped by a machine gun." The unit then pulled back and had to leave Staff Sergeant Buck and one other soldier's bodies due to the tremendous volume of incoming fire. A Korean War veteran, Buck left behind a wife and seven children.

Doc Shyab stated, "I talked to Sergeant Buck's radioman and he said, 'I felt horrible leaving him there but he was dead.' During the retreat to safety, Captain Konnerman's radioman, Truman Lovins, was shot in the arm, which I fixed the best I could. Roy Allen Lamon was machine gunned down during the retreat and was also left behind. The constant pounding from the artillery, gunships, and air strikes should get the enemy to retreat this area.

"However, we found our enemy had not moved one inch and we got viciously hammered again on our next assault up the hill to get our dead. There were hidden enemy bunkers everywhere, no matter which direction we took up the hill.

"Every man in C Company is firing back with all of our weapons but we still could not get this enemy force to withdraw in the fight. Our company had taken on many more wounded and dead again. We attacked, retreated again to bomb this place full of enemy soldiers, and then attacked again.

"For some strange reason, Louie Kimbrell wanted to go with each advancing platoon up to fight. Louie Kimbrell, the brave young soldier he was, did more than his share of killing on Chu Moor. He was one of the bravest soldiers ever and will never be forgotten by the men of C Company."

Casualties on both sides continued to increase as the day progressed into the evening. It was now estimated that the enemy force was at least a battalion and perhaps regimental in size. Heroism was a common occurrence as evidenced by the day's body count for 1/22: 53 WIAs and seven KIAs (of which two were initially carried as MIAs). There were nine confirmed NVA killed. The day's support included nine gunships, fourteen air strikes, 1,622 artillery, and 218 mortar rounds fired.

"The evening of the 27th our CO (Captain Konnerman) was giving a sit rep (situation report) for our company and was talking to a major at the base camp telling him about the condition of our company's ability to attack the NVA in light of the many casualties we had taken," reported Doc Shyab. "The captain raised the question about another attack." The Major replied, 'I don't care how many men you have left, attack the hill!' These words I heard as I was very near the RTO when the reply came through. All who heard it figured that we were going up the hill to die."

Elmer Hale, who was the RTO for Captain Konnerman asked him, "Is this damn hill worth these men's lives?" He replied, "I agree with you, but I've got to try. If not, I'm going to lose my bars." "He had no control over the matter. He was just following orders the same as I was following orders."

In the northern sector of the valley, B and C/1/14 conducted helicopter combat assaults in support of 1/22. C Company killed one NVA while A/1/14 located enemy tunnels and recovered some partially burned documents.

DAILY STAFF JOURNAL OR DUTY OFFICER'S LOG

ORGANIZATION OR INSTALLATION		S-3, 1st Bn 14th Infantry GOLDEN DRAGONS
PERIOD COVERED		0001, 27 Apr 68 - 2400 27 Apr 68
Item No.	Time	Incidents, Messages, Orders, etc.
	1600	C/1/14 has one VNA KIA at loc 838913. Ind was wearing Unif w/steel pot.
		SUMMARY: GOLDEN DRAGONS continued operations in assigned AO. Alpha Co found 2 tunnels and some partly burned documents. Documents found to be NVA clinging records. Bravo and Charlie Co's conducted CA in support of 1/22 Inf. Charlie Co had small contact with NVA. Results 1 NVA KIA. C Company OPCON to 1/22 Inf. PLANS: Conduct CA with Alpha Co to support ground operations of the 1/22 Inf. Delta Co to remain as security for LZ SWINGER.

4TH INFANTRY DIVISION
OPERATIONS SUMMARY
21 APR 68—30 APR 68

27 April. At YA842884 Company C, 1st Battalion, 22nd Infantry received one B-40 rocket round. At YA845893, the night location for Companies A and B, a patrol from Company A received SA and B-40 rocket fire 30 meters outside their perimeter. The two companies remained in their night location and continue to receive sporadic SA, rocket and mortar fire throughout the day. At 0940 hours Company C conducted an attack against an unknown size force in bunkers. Meeting heavy resistance, the company withdrew and called in artillery and airstrikes. At 1040 hours, the heavy enemy fire forced them to withdraw to the night location. Artillery and airstrikes continued to hit the enemy position and at 1615 hours Company C took the high ground and the bunker complex. At 1700 hours the enemy counterattacked with AW, B-40 rocket and 60mm mortar fire. At 1825 hours, the company withdrew, but continued to receive mortar fire. At 1920 hours artillery produced a large

secondary explosion at YA835819 and the mortar fire ceased. Results of the contact were five US KIA, 53 WIA, two MIA, and nine known NVA KIA. At YA934944 Company C, 2nd Battalion, 35th Infantry found one NVA body. At YA839911 Company C, 1st Battalion, 12th Infantry killed one NVA. The airstrip at BAN ME THOUT received 40-60 rounds of 82mm mortar fire, resulting in 14 US WIA and seven aircraft damaged.

CHAPTER 9
Day 5: Sunday, 28 April 1968

"I saw the medics with soldiers leaning on the big tree, putting IV's in their arms, and checking their eyes for signs of life."

DAILY STAFF JOURNAL OR DUTY OFFICER'S LOG

ORGANIZATION OR INSTALLATION		S-3, 1st Bn 22nd Infantry REGULARS
LOCATION		FSB YA868888
PERIOD COVERED		0001 28 Apr 68 - 2400 28 Apr 68
Item No.	Time	Incidents, Messages, Orders, etc.
	0740	C/1/14 in contact.
	0837	Air strikes employed (mixed ordnance).
	0910	B Co recon by fire to N.
	0917	A Co recon by fire and received small arms fire in return.
	0940	C Co in contact with small arms.
	0943	C/1/14 slight contact.
	1020	C Co receiving incoming mortar rounds.
	1033	Air strikes employed (mixed ordnance).
	1040	C Co receiving mortars (very accurate fire).
	1115	C Co receiving mortar rounds.
	1117	Gunships check in (Gambler 9).
	1120	C Co receiving mortar rounds.
	1203	B/1/14 OPCON to 1/22.

	1235	A and C 1/14 linked up.
	1251	C Co received 13 mortar rounds inside their perimeter.
	1334	B/1/14 being flanked by enemy.
	1344	Air strikes employed (mixed ordnance).
	1347	B/1/14 spotted 7-10 enemy.
	1422	C Co receiving incoming mortar rounds.
	1510	B Co receiving incoming B-40 rockets.
	1550	C Co receiving incoming mortar rounds.
	1633	C Co extracted: D/1/14 inserted.
	1640	B Co recon by fire.
	1715	Air strikes employed (mixed ordnance).
	1748	B Co receiving mortar rounds.
	1802	D/1/14 receiving mortar rounds.
	1924	FSB received mortar rounds.
		The casualties for the day included 37 WIA's, 1 MIA, 7 KIA's. Support consisted of 9 gunships, 10 air strikes, 1,521 artillery rounds and 499 mortar rounds.

Specialist 4 Sam Drake: "I don't think we moved up the hill because we were too tired and beat up with half of our Company wounded or killed. I don't remember exactly what happened that morning because it was extremely stressful! We started reinforcing our bunker with more sand bags for the next round. Richard Cassano and I shared the same bunker along with Sergeant Cappaletti. I remember having Richard holding the sand bag while I shoveled dirt into the bag. Richard had his feet in the bunker entrance fearing all the time that we will be hit with mortars. That's how bad our situation had become. It was getting harder to find soft dirt so I told Richard he was going to have to leave the bunker entrance and help hold the bag. All the time Sergeant Cappaletti was walking around in a daze. I think he was shell-shocked. Within the next few minutes Sergeant Cap was being loaded up in the helicopter and headed for the firebase. It was becoming extremely bad!"

Drake continues, "The helicopter was bringing in more troops from the 1/14. The 1/14 supplied more body bags for the dead that they left with our wounded. Some of the troops that got off didn't know that our LZ was hot. They just walked up the hill from the LZ. The NVA were mortaring us constantly. Lieutenant Zimmerman was on one of the helicopters that came in and I remember some of his guys greeting him at his 3rd platoon bunker. Unfortunately, his 3rd platoon was zeroed in by the NVA. I was on the opposite side of the big tree when the NVA mortar hit the 3rd platoon bunker. I was hit in my left hand and dove to the base of the large tree. Richard Cassano was near the 3rd platoon bunker when he was hit. After being hit he ran to our side of the tree; he fell short so we pulled him to the base of the tree. Bud Roach tried to save him; however he couldn't."

Lieutenant John Bobb continues, "April 28 became worse. Hard to believe, but it did. My platoon was in the point, 10 to 2 position in the defensive perimeter, facing the hill we were assaulting. 2nd platoon was in the 2 to 6 position, and 3rd platoon, was in the 6 to 10 position. The second platoon was on the side with the small makeshift LZ, and facing the valley towards the battalion firebase. The 3rd platoon faced the valley, and mountains towards enemy territory.

"A number of things happened on this day. We were sent a 90mm recoilless rifle, with flechette rounds. I gave it to Louis Kimbrell. It appeared the enemy was starting to probe our perimeter. I had to worry about the back blast into the perimeter. Everyone got down. I would load Louis up, check to the rear, and slap his helmet. And he would fire into the bushes and trees. I would reload him and start the procedure over again. We were not attacked in force. They didn't need to.

"In the rear, Lieutenant Zimmerman about to go to Hawaii learned of our predicament and the platoon he commanded for eight months being in trouble. The senior medic, Charles Shyab, had been severely wounded and dusted off. Former senior medic and 3rd platoon medic, Bud Roach, and Zimmerman volunteered to rejoin their old outfit.

"I was in the perimeter, and one of Lieutenant Zimmerman's soldier's, Doug Royalty, came over to me. He showed me a photo of his wife and children. He said he wanted to show it to someone because Lieutenant Zimmerman wasn't there. He told me he knew none of us were going to get off that hill alive, and that he felt better knowing he had shown me the photo of his family before he died. I didn't want to tell him that a bunch of us had just decided to kiss our asses' goodbye before he approached me with the same conclusion. Weird feelings to know you are already dead, and you just keep doing what you are told anyway.

"I had no training or education in these matters. Like when a few months earlier, I was a new Lt running basic trainees at Ft. Lewis, Washington. We were out on maneuvers, and in the middle of the night I had to roust some young soldier out of his hooch, bring him back to the main road, where a deuce and a half truck was waiting; told him his family house just burned down, killing his parents, and all of his sisters. He got in the back of the truck, and it left. I had no damn idea what I was doing. I had people older, more mature and more experienced, looking to me for some sort of comfort. I didn't know how to give it.

"A helicopter came in with Lieutenant Bill Zimmerman. After Bill talked to the Captain, he and I met in the middle of the perimeter. We crouched down and talked. I have no idea about what. It was my platoon's turn to attack the hill on point. I remember Bill moving among his men. His platoon area was just being devastated by enemy indirect weapons. I remember looking over there again while I was telling my men to saddle up. Meaning get out of the bunkers and get in formation to move up the hill. Everybody can think my next observation is bullshit, I don't care. I looked towards the third platoon and saw Lieutenant Bill and Royalty filling up sandbags to refortify their position. I turned back right and started saying something to Danny Brooks. I saw something blur my vision between Danny and myself. It was a mortar round. When it was noisy, i.e. choppers etc., you couldn't hear the NVA mortar tubes.

"Mortars are an indirect fire weapon unlike a rifle or shoulder fired rocket. They make a very unique hollow thumping sound when the round leaves the mortar tube. When you hear it, you have time to roll into a bunker and get down before it lands. The operative phrase, is 'when, you hear it.' Throw in medevac dust off choppers, supporting fire, rockets, bombs, and other loud noises, when you are assembling to assault the hill, and you are standing straight up in the open when the mortar rounds land, because you didn't hear the 'when'.

"I do not recall the number of salvos the following took, or maybe it was only one or two. I have no idea. It seems the whole day went like this. The end result was generally this; I was blown backwards, it wasn't the only time up there that I had been knocked totally senseless. I don't know how much time passed. Probably not much. While lying on my back, I felt blood on my jungle fatigues. I unbuttoned them to see if it was mine or someone else's. I seemed OK. I got up and looked around. Lieutenant Zimmerman and Doug Royalty were dead. The entire Charlie Company Command structure was wounded. The first sergeant, the artillery Lieutenant, Jack Chavez, the mortar forward observer was wounded in the face shielding the company commander, Captain Konnerman. I remember seeing the Captain with his shirt off, with two holes in his chest. Blood was dripping out of them.

"I started checking on my men. Danny Brooks was lying down with one of his forearms sliced to the bone by shrapnel. One of my soldiers was walking towards me with his shirt open, totally in shock. He had four huge shrapnel holes in his chest. Everybody had to be directed or helped towards the LZ to be dusted off. Other soldiers were lying around this famous tree we all remember, with medics putting IV blood expander units in their arms, giving some morphine, and checking their eyes as the life slowly drained out of America's finest. Due to the casualties in the 3rd platoon, I had to detail a couple of my soldiers to their side of the perimeter, which was ground zero every time we got hit. I checked on them. Every 3rd platoon bunker had a dead soldier in the bottom of it, as if they were sleeping.

"I saw the medics with soldiers leaning on the big tree, putting IV's in their arms, and checking their eyes for signs of life.

"I think we then tried to make the assault we had intended on, and I remember getting in the clearing in front of my point bunker, and mortars landing again. I have a faint memory of being blown up in the air, again without a scratch. It was at about this point where we didn't have enough men left to secure the perimeter, let alone assault the hill.

"I don't know much, but I do know if we would have all just run up that hill at once with no secure LZ, and no perimeter, we would have been 'effed. We were vastly outnumbered.

"Senior medic Bud Roach, who volunteered to come back, ended up putting his old unit that he had served with in military body bags for transport off the hill.

"The collective memories of many of Charlie Company survivors are that we had 17 soldiers left on that hill out of about 110 or 120. The Captain was standing there upright and in the open with a radio in each ear, being assisted by Elmer Hale. He was taking directions from battalion, giving them information, and calling in supporting fire and F-4's on the hills where the mortar fire was coming from. When we would get secondary explosions that meant we had taken out one of the NVA positions. We would see the secondary explosions and get happy; whatever the degree of happy was.

"Later that day, we were evidently going to be replaced by a unit of the 14th or 35th Infantry. We did not have enough soldiers left to man the perimeter, let alone assault the hill. It is my general understanding that the unit that replaced us got all chopped up also, and it took a B-52 strike (arc light) and a sweep of the area by the 4th Cavalry to recover all of the dead soldiers, including mine.

"I have only a few memories before leaving. I remember sitting with Pappy's rucksack between my legs.

"Then I remember standing with my back to the hill, in the open. I felt like I was in the middle of a vampire movie. I seem to remember there was smoke or fog all around me; like nothing was real. I remember seeing an American Soldier walking towards me. He wasn't in our company. I just stood there, and he kept getting closer and closer. It is all I remember."

Just prior to the Chu Moor battle, one of C Company's platoon leaders, 1st Lieutenant William Zimmerman, left the field to become the new Executive Officer. First though, he planned to meet his wife in Hawaii for an R and R vacation. When it was learned that Zimmerman's replacement had been wounded, it was requested that he return to the field to help out with the dangerous situation on the mountain. On the morning of 28 April he jumped off the helicopter into C Company's position, now located on the lower, southern crest. Despite heavy automatic weapons fire, along with exploding mortars and rockets, he rushed to the perimeter positions and helped to reorganize and inspire his troops. Within twenty minutes on the ground, Lieutenant Zimmerman was mortally wounded by a hostile mortar round. For his outstanding leadership and aggressive actions, he was posthumously awarded the Bronze Star. Sadly, his wife learned of his death just before boarding the plane that would have taken her to Hawaii.

During the same mortar attack that killed Lieutenant Zimmerman, C Company's FO, Jack Chavez was blasted into the air and landed on some dead tree roots, covered in blood. He was able to crawl into a bunker, looking like hell, but still alive. Captain Larry Konnerman was also wounded from another mortar round as bodies fell everywhere.

Medic Charles Shyab remembered, "Immediately after an NVA probe on the morning of the 28th they mortared us at the exact spot where they had probed us earlier. I gave aid to an RTO whose last name was Askew, who was attached to Lieutenant Zimmerman who was KIA from the mortar. Sometime during mid-morning they found our bunker perimeter. We received the order to move up the hill again as our planes dropped ordnance on the hill, prepping for our attack. They used the noise of our jets to cover their mortar attack. Being the senior medic there was no way I would not be there to help my buddies even though my life was at risk. I had crawled out of my bunker to join in on the attack as the plane flew over."

Specialist-4 Charles Shyab was a Seventh-day Adventist conscientious objector. Earlier in the battle he had lost both his helmet and glasses and by not carrying a weapon, he was left totally defenseless. "On the morning of the 28th of April when the situation was deteriorating I read from the New Testament (he still has his blood stained

Bible to this day) and prayed to the Lord, telling Him that there was no way I could survive the danger unless He preserved my life.

"I promised Him that if I was to survive that I would do what He wanted me to do and if He wanted I would be an Adventist Teacher."

Shyab recalls, "When the command came for us to leave our bunkers to attack the hill a jet was dropping a 500 pound bomb on the mountain. The NVA used the sound to cover the mortar noise as we were moving up. About 20 feet away I saw a flash and felt my right hand go numb just like you do when you bump your elbow. I thought I had just lost my hand but I looked down and it was still there; I knew I was hit so I returned to the bunker and told the 1st Sergeant I was hit. He lifted my right collar and his eyes got big as he confirmed the injury. I was also hit in the other arm and both legs but I did not know that. The medevac chopper came in, with three gunships to cover us and suppress any NVA fire. Of the 120 men in the company, 30 were KIA and 15 were choppered out; the 70 injured were treated by their fellow soldiers or medics.

"My friend Richard Cassano escorted me to the LZ to help load the wounded; he never made it back to his foxhole!

"Another great medic, Bud Roach, came in to replace me when I left. Come to find out that there were no other medics on the ground and that I was the last one to leave till he arrived to evacuate the rest of the men.

"When the chopper landed at the aid station the corpsmen did not move because we were such a sight; me with no helmet, just a boonie hat, sunglasses, a five-day beard, and a messy uniform. The sergeant ordered them to get the wounded off because they were hesitant to be in such a mess. Inside the station they thought I was a hippy and moved me to a corner to protest the war!"

Doc Shyab fulfilled his promise to God and became an Adventist teacher and school principal. According to more than one soldier, Shyab was revered by the men in his company.

At LZ C-Rations, Specialist-4 Bud Roach recalled, "On 28 April a call came in that C Company medics were wounded and they needed replacements. I agreed to return to my old unit as did Bill Zimmerman who had been my platoon leader with C Company. I was given body bags and told to get the wounded and KIA's out and that C Company was being removed.

"Bill went in on a helicopter just a few minutes before I got a flight. I rode in on a chopper carrying a re-supply of ammo. I was on my knees between the pilots' seats so I could see our destination clearly. Was I scared? For a long time I didn't want to admit just how scared I was. I think I can relate to the soldiers who were on the landing craft approaching Omaha Beach in World War II. I knew the situation I was headed for and had time to think about it. My heart was racing. I was so excited that I could taste the adrenalin.

"As the helicopter neared the landing zone, shots were being fired at it. The dull thud of the bullets hitting the chopper is an unmistakable sound. I can't say it enough; helicopter pilots had ice water running through their veins. They did this time and time again. The landing zone was a little patch of chopped down trees on the side of the mountain. The chopper hardly touched down. When I got on the ground I was shocked at what I saw.

"C Company was on the lower crest. The chopper came in from the east to the west into a makeshift LZ that was still littered with felled trees. Some things have faded in my memory but the memory of that first sight on Chu Moor is as clear as a snapshot. There was a bunker on the left where two soldiers were working on a wounded GI on top of it. I went to check the wounded person and found that he was KIA. I made the soldiers get into the bunker. Directly in front of me was a huge tree with a bunker to the right of it. On past the bunker was a small clearing with a trail curving slightly to the right and up the ridge to the higher peak. The clearing was a kill zone. The NVA had it zeroed in for mortars and rockets. When a chopper came in, the NVA would open up with mortars under cover of the noise from the choppers.

"In the clearing were the bodies of two KIAs. The First Sergeant, Richard Cassano, and I went out to retrieve the bodies of 1st Lieutenant William Zimmerman and Specialist-4 Amel Royalty. While we were doing this, a chopper was approaching. Knowing the NVA would fire mortars, we were running around the big tree to get to a bunker when a mortar round exploded. Cassano behind me was killed and the First Sergeant in front of me was seriously wounded. I did not get a scratch. The big tree saved my life. I turned to Cassano and he was obviously in trouble. I leaned him against the tree and tore his shirt open. He had a very large hole in his chest and was losing color. I knew I was losing him. The wound was too serious, I couldn't save him. I turned to the First Sergeant and patched him up until a chopper could carry him out."

Roach continues, "C Company began being replaced by another unit. A chopper would bring in about six from the new unit and about six from C Company would leave. I went out late with the KIAs back to the firebase. During the last few years I have talked to the Medical Officer, Bill French, who was at the firebase. He said when I got off the chopper the only thing I said was, '*They're all dead.*' I don't remember saying anything. I went to the aid station bunker to be by myself. Just about every friend I had in Nam had been wounded or killed. It was never the same."

Sam Drake remembers C Company being pulled out and being replaced with the 1/14, "I do remember getting on the helicopter and hoping we would clear the area before the incoming mortars hit. It was a short ride to the firebase. I do remember landing and seeing all the body bags lying around the LZ with our guys inside them. I didn't remember too much after that, I think I was in shock."

Back at LZ C-Rations, Lt Bobb recalls, "The next thing I remember, I was standing on the chopper landing pad at Chu Ya Bruh, (LZ C-Rations) with the battalion first sergeant or another senior NCO yelling at everybody to clear their weapons before leaving the Chopper pad. He was probably afraid the Battalion Tactical Operations Center would be the object of our next assauLieutenant

"Once inside the perimeter I observed an NCO looking at the hill we had just come from through a pair of artillery spotting scope binoculars on a tripod. He turned towards me and said he could see the NVA in the trees shooting at us. I didn't say anything, but I was thinking,

'Hey, thanks for the info, too bad I didn't have it, before the first time we went up the hill, three days ago.'

"We would normally be required to take our place on the perimeter, and be required to dig in. I just remember lying on a nice grassy slope within the perimeter. I rolled up in my poncho and poncho liner and laid there. That night I heard an enemy mortar tube go off several times in a row. This meant that instead of taking time to bracket us for more accurate fire, they opted to use Kentucky elevation, and Tennessee windage to throw all the rounds down real fast and then move before we returned fire. I already knew that all the rounds would land in the same place, so I didn't move. The first round went long over our heads, so I knew the rest would also. I just laid there. I then heard a loud explosion within the perimeter, very close. I still didn't move. I found out the next morning, that D Company, now down from the hill, and in the perimeter, set up a radio antenna in front of a 105mm howitzer. They got a fire mission in the middle of the night, and a round hit the antenna, killing and wounding more of D Company.

"I was sitting on that same grassy slope with the remaining members of Charlie Company. I seem to remember there were 26 of us. Some of the more minor injuries had evidently already returned. I think I was still in the same bloody clothes, just sitting there staring at nothing.

"A chopper lands, and a very clean Lt jumps off the chopper, with a swagger and energy of a clean well-fed and well-conditioned Lieutenant Now I realize how stupid I looked when Roark saw me coming. He bounded up to our little group, all happy to be there. It turns out he was Lieutenant Zimmerman's 3rd platoon replacement. He stood in front of the little group, all excited, and enthusiastic, and asked, 'Which platoon of Charlie Company is this?' A voice behind me, I didn't turn to look; said in a low dead pan voice, 'This is Charlie Company'.

"Then the Lieutenant took a good look at us, and his facial expression changed. He turned out to be a real good guy.

"Another memory is a captain came from battalion and told me that a 105 howitzer and a bunch of ammo being transported under a Chinook (Shithook) helicopter had fallen off, and was on the ridgeline outside the perimeter, in the opposite direction of Chu Moor. I was directed to take the remaining members of Charlie Company and escort an artillery captain to the site of the downed artillery piece so he could disable it, so the enemy could not use it.

"We moved down the ridgeline very fast. The artillery captain didn't like it. He was probably right; but I didn't give a shit. A little later I found out why he was right.

"We reached the site of the howitzer which fell barrel first into the ground with such force that the trails (the two back support arms) bent. The ammo was scattered around. I set up a security perimeter while the artillery captain used thermite grenades to weld moveable parts of the breach block to prevent enemy use. He then stacked up all the ammo around the howitzer and began to wire up blocks of C-4 explosive and non-electric blasting caps and time fuse.

"When the captain was almost done, I took a couple of spare thermite grenades and walked over to the edge of the perimeter, and found some very dry clumps of bamboo and brush. I looked over the valley and got the bright idea to throw the thermite grenades into the dry vegetation and just start a fire and hope the entire country of Vietnam would just burn down. It was not an authorized or well thought out action; but again, I didn't give a shit. The thermite grenades fizzled out, and I shrugged my shoulders, and walked back into the perimeter.

"I found out later that Wayne Raymond and his slinky recon types were lurking around; and if I had burned Wayne, he wouldn't be my friend today.

"Now, I found out why the Captain was concerned about our rate of descent. He was looking at his watch on the way down, because it is the standard he was going to use to safely set the time fuse, so that we were well within the perimeter before the C-4 and all the H. E. artillery rounds detonated, sending shrapnel all over at several thousand feet per second. I don't know if any of you have ever seen shrapnel from military ordnance in combat; but it is jagged, irregular shaped, in varying sizes, and does not have an independent thought process to decide who to kill, or distinguish between friendly and enemy troops. I don't know if this technical glitch has been solved to this day.

"There are just certain rules that don't change. You can't move uphill as fast as you can downhill. The captain lit the fuse, we moved uphill. About three-quarters of the way up the hill, the captain told me we had a short time before there was going to be a very large and dangerous explosion. I instructed my men to find a tree hopefully wider than their bodies, hunker down on the uphill side of it, and wait. I did the same, as did the captain.

"The captain's timing as to when the explosion was going to occur was right on. It was ugly, and copious amounts of shrapnel found its way to our location at a rapid pace. We could hear it. We could hear a lot of it. Nobody was hit. What is the old saying, *'God takes care of fools and crazy people'.*"

Sergeant Bliefernicht (B/1/22) recalls, "On the morning of April 28, I was asked to pull security to the rear side of our company. Another trooper and I were walking towards our positions for security. We were about ten to fifteen yards apart when a stray 105 shell landed between the two of us. He was killed instantly, and I was blown back a good ten yards. I was peppered with shrapnel. The medic came and started to examine me, I had a lot of blood coming from between my legs. First question I asked the medic was are my family jewels OK? He said no problem at all for my family jewels. He bandaged my left arm up. I later learned that another trooper to the left of us was also killed by the same shell. I never learned the names of the two troopers killed by that 105 shell, they were from different units.

"Right after I was treated for my wounds I went into shock of some kind. I remember clearly of getting up and walking around and saying I don't care anymore, or words to that effect. Bullets were flying all around me; somebody grabbed me, and pulled me down. I got up again and started walking around again. Sergeant Cowley grabbed me and tied me to the backside of a big tree. A little later on I was loaded on a chopper with one other trooper that was wounded.

"Also loaded on the chopper were three of our K.I.A.'s. During the trip to Pleiku, money and MPC was floating around the chopper's interior from one of the dead troopers. The door gunner picked up the money and gave it to me saying, it belongs to you guys. The money was all soaked with blood; I offered the money to the trooper next to me. He refused to take the money.

I threw the money out of the chopper and I said, 'It doesn't mean nothin'.

"I recovered from my wounds at Cam Ranh Bay. When I returned to my company, I was told there was an inquiry of why there were stray 105s coming into our area. It was from another firebase firing a worn out 105. The officer in charge knew the 105 was firing inaccurately, but kept firing, no charges were filed."

B/1/14 was operating in the valley to the northwest of Chu Moor Mountain in an effort to stem any enemy reinforcements from joining in on the mountain fight.

As A and C 1/14 began climbing up Chu Moor from the north, they, too, encountered the enemy and had nine wounded and two killed, while they inflicted five deaths upon the enemy.

While conducting a combat assault onto the mountain to relieve C/1/22, D/1/14 received mortar rounds on their landing zone that resulted in five friendly wounded casualties. B Company, 1/8 had relieved D/1/14 as defense security for LZ Swinger.

LZ Swinger received incoming enemy mortar rounds without any damage or injuries. They promptly called in artillery support against the suspected mortar positions that resulted in one large secondary explosion.

The casualties for the day included 37 WIA's, one MIA and seven KIA's. Support consisted of nine gunships, ten air strikes, 1,521 artillery rounds, and 499 mortar rounds.

DAILY STAFF JOURNAL OR DUTY OFFICER'S LOG

ORGANIZATION OR INSTALLATION		S-3, 1st Bn 14th Infantry GOLDEN DRAGONS
LOCATION		
PERIOD COVERED		0001 28 Apr 68 - 2400 28 Apr 68
Item No.	Time	Incidents, Messages, Orders, etc.
	1555	D/1/14 had two WIA at loc 845904. Dustoff completed at 1555 hrs. Extent of wounds unk at this time. 1 sprained ankle & 1 cut hand.
	1906	D/1/14 recd 5 incoming mortar rds neg casualties.
	1920	C/1/14 7 WIA & 1 KIA. Loc 849897.
		SUMMARY: GOLDEN DRAGONS continued operations in support of 1/22 Inf. Alpha Co conducted CA near Charlie Co 1/14 and began to conduct operations with Charlie. Alpha and Bravo Co's revert OPCON to 1/22. Charlie Co while OPCON had 5 NVA KIA and 1 wpn (AK-47) CIA. Delta also OPCON to 1/22. During CA Delta Co received mortar fire on LA, results 5 WIA. LZ SWINGER received 7 incoming mortar rounds. AS was called and results one large secondary explosion. 1/14 receives OPCON Bravo Co 1/8 Inf for security of LZ SWINGER. PLANS: Elements of 1/14 to continue support of operations of 1/22 Inf.

4TH INFANTRY DIVISION
OPERATIONS SUMMARY
21 APR 68—30 APR 68

28 April. At YA844891 a 105mm artillery round impacted inside the perimeter of Company C, 1st Battalion, 22nd Infantry, resulting in one US killed, and one US wounded. At YA842907 Company C, 1st Battalion, 14th Infantry engaged five NVA, killing four and capturing an AK-47, with one US slightly wounded. At YA843887 the company again made contact and withdrew to call in artillery and airstrikes. Half an hour later the company received a 60mm mortar attack, resulting in

four US KIA and 19 US WIA. At 1424 hours Company D, 1st Battalion, 14th Infantry attempted a lift into Company C's location but the mission was aborted by mortar fire on the LZ, wounding three US. At 1700 hours Company D received three rounds of 60mm mortar fire at YA842884, resulting in two US WIA. At 1906 hours at the same location the company received five 75mm RR rounds, resulting in one US KIA seven US WIA. At YB830183 a mine sweep element from Company D, 3rd Battalion, and 8th Infantry found one NVA body at YA849168, and Company A found three NVA bodies at YB830183. At ZA072769 Troop A, 1st Squadron, 10th Cavalry found one NVA body. At YA870880 the Commanding Officer of Company A, 3rd Battalion, 12th Infantry moved off the trail his unit was moving on and was captured by two NVA.

CHAPTER 10
Day 6: Monday, 29 April 1968

"The firebase was crowded and alive with activity."

DAILY STAFF JOURNAL OR DUTY OFFICER'S LOG

ORGANIZATION OR INSTALLATION		S-3, 1st Bn 14th Infantry GOLDEN DRAGONS
LOCATION		FSB YA868888
PERIOD COVERED		0001 29 Apr 68 - 2400 29 Apr 68
Item No.	Time	Incidents, Messages, Orders, etc.
	0040	A and B 1/22 have movement, recon by fire.
	0225	B/1/14 receiving 60mm mortars, 82mm mortars, B-40's, recoilless rifle and flame throwers.
	0300	Spooky 22 checked in.
	0440	Spooky 23 checked in.
	0635	Air strikes employed (mixed ordnance).
	0647	C/1/14 has movement to south
	0705	Air strikes employed (mixed ordnance).
	1030	Strobe jet took pictures.
	1141	B/1/14 returned to parent unit.
	1201	B/1/22 received 2 B-40 rockets. Company B/1/22 received B-40 rockets within their perimeter on 29 April that resulted in 3 wounded soldiers.
	1225	D/1/14 received incoming mortar rounds.

	1415	Air strikes employed (mixed ordnance).
	1459	B/1/22 receiving incoming mortars and small arms fire.
	1510	Air strikes employed (mixed ordnance).
	1637	A and C 1/14 returned to parent unit
	1645	Air strikes employed (mixed ordnance).
	1714	Gunships checked in (Crocodile 6).

B/1/14, still in the northwest area of operations, encountered 60mm and 82mm mortars, B-40 rockets, and flame throwers coming in from the east of their position. They initially had one KIA and five WIAs during the action. It was estimated that they had hit an enemy force of a battalion size or larger. The NVA were dressed in tan uniforms and had new web equipment and packs. The enemy made a strong ground attack against their defensive position. Using all of the available support assets, it had a confirmed body count of 46 NVA and it was estimated that an additional 150 were killed by air support. B Company suffered two killed and eleven wounded by the end of the battle.

Company B/1/22 received B-40 rockets within their perimeter on 29 April that resulted in three wounded soldiers.

By the early afternoon, the 14th Infantry units were ordered to return to LZ Swinger, which would be totally evacuated within the next few days. As D/1/14 was being extracted from the mountain by helicopters, incoming mortar rounds wounded 13 soldiers and killed one, thus giving the NVA last blood. B/1/8 was relieved on LZ Swinger and returned to their parent unit.

At 1811, A and B 1/22 headed back to LZ- C-Rations. Gunships (Cougar 13) checked in and provided escort for the two depleted companies. The total casualties during the day's fighting were 40 WIA's and five KIA's. Seven gunships and 12 air strikes were employed for protection.

The commander of Company A/1/12, who had been captured the day before, managed to escape his captors when an artillery round wounded his two NVA guards when the round struck close to their position. He was able to reach LZ C-Rations unharmed.

Bud Roach stated, "April 29 and 30 were two busy days at the firebase. Most of the units engaged on Chu Moor were brought to the firebase and then carried by Chinooks back to the base camp to be re-fitted because of the high rate of casualties in the battle. Everyone was called in. The firebase was crowded and alive with activity."

DAILY STAFF JOURNAL OR DUTY OFFICER'S LOG

ORGANIZATION OR INSTALLATION		S-3, 1st Bn 14th Infantry GOLDEN DRAGONS
LOCATION		
PERIOD COVERED		0001 29 Apr 68 - 2400 29 Apr 68
Item No.	Time	Incidents, Messages, Orders, etc.
	0405	B/1/14 while OPCON to 1/22 report received 60mm, 82mm B-40 rockets and flame throwers. Estimate fire coming from the East believe to Bn or larger. Results 1 US KIA, 5 US WIA.
	0510	Total 2 US KIA & 11 WIA, location 828908. B/1/14 while OPCON 1/22 its element is receiving small arms fire from the west. Returned fire results 1 NVA KIA & 1 SKS location 828908.
	0800	B/1/14 while OPCON to 1/22 has body count (est.) of 150 NVA KIA. Location 828908
	0815	1, B/1/14 is in contact at location 828908. 3, location 83910 has 40 NVA KIA's 8 weapons, 1 B-40 Launcher, 2 heavy MG, 3 AK-47s, 2 SKS's, numerous rockets, hand grenades, Ind were dressed I K unit w/new web gear & packs.
	0848	B/1/14 identified 2 NVA units as 507th 3/7 Inf.
	1230	Gladiator dustoff completed dustoff for B/1/14 6 WIA's wounded in last night action. LZ Swinger. Completed at 1300 hrs. Also requested dustoff for 4 more of B/1/14 WIA's comp.
	1305	D/1/14 received 3 incoming mortar rounds negative casualties at this time. Results 13 US WIA's 1 US KIA.

	2105	Be prepared to move 2 companies from LZ SWINGER to 912958 (Hill 1042) at 1000 hr. 30 April. The remainder of the 1/14 and 2/9 Arty will move to POLEI KLENG sometime the same day. Fire Base Singer will be completely cleared.
		SUMMARY: GOLDEN DRAGONS saw Bravo Co while OPCON to 1/22 Inf. had ground attack on the defensive position. Results 46 NVA KIA (B.C.) 150 NVA KIA (EST) 8 weapons CIA, 2 heavy MG CIA, numerous rockets and hand grenades CIA. Ind. NVA were dressed in tan uniforms with new web equipment and packs. All units returned to 1/14 Inf. Bravo reverted OPCON to parent unit. During extraction of Delta Co mortar round were receive on PZ. Results 13 US WIA, 1 US KIA. Alpha Co and Bravo Co returned to original position upon return to LZ SWINGER. Charlie and Delta Co's to secure LS SWINGER. PLANS: 1/14 to move from LZ SWINGER to establish new FSB vicinity coordinates 899933. New FSB to be named LZ ROBERTS.

4TH INFANTRY DIVISION
OPERATIONS SUMMARY
21 APR 68—30 APR 68

29 April. At 0206 HOURS AT ya836908 The night position of Company B, 1st Battalion, 14th Infantry was attacked by an estimated NVA battalion employing SA, AW and B-40 rocket fire. At 0224 hours the enemy began employing 60mm and 82mm mortar fire and flame throwers. The enemy began digging in 10-15 meters outside of the company perimeter, and heavy contact continued until 0515 hours when the enemy broke off and withdrew. Results of the were two US KIA, 20 US WIA, 46 NVA KIA in the immediate vicinity of the perimeter, two light machine guns, four AK-47's, two SKS and one B-40 rocket launcher captured. AT ZA012813 one US from Company D, 1st Battalion, 12th Infantry was wounded in an exchange of fire with two NVA. At YB928202 Company B, 3d Battalion, 12th Infantry found one NVA body in a grave. At YA844892 Company B, 1st Battalion, 22d Infantry received one B-40 rocket, resulting in three US WIA. At ZA063283 an

airborne personnel detector received ground fire, resulting in one US WIA. At YA936935 Company B, 2d Battalion, 35th Infantry found two NVA bodies. AT ZA115513 an APC from Troop B, 1st Squadron, 10th Cavalry received two B-40 rockets, resulting in one US WIA. The Commanding Officer of Company A, 3d Battalion, 12th Infantry captured on 28 April, returned to the 1st Battalion, 22d Infantry FSB. He had made his escape when an artillery round wounded his two NVA captors at YA878875.

CHAPTER 11
Day 7: Tuesday, 30 April 1968

"The soldiers had proudly stood toe to toe and held their own against a larger, well entrenched, superior force on the now forgotten mountain."

ON LZ C-RATIONS, 1/22 ALSO MADE PREPARATIONS TO ABANDON their base. The plans were to pull all U.S. Forces out of the area and call in B-52 strikes, called arc light strikes. The firebase was a busy place, being the staging point of everyone leaving the area. It was crowded.

Roach recalls, "On the night of April 30 an artillery round exploded as it left the barrel of the cannon. The shrapnel wounded several in the firebase. We medics were called out in the darkness to do our job. There was a lot of disorganization because nobody knew what had happened. We called for a dustoff for six of the most seriously wounded. The pilots did not want to fly in the hazardous conditions at night but after some discussion about who would be responsible if the wounded died, the choppers came and the injured were medevac'd to a field hospital."

The outgoing 105mm round inadvertently struck a radio antenna that resulted in an airburst that killed one and wounded 12 soldiers. This was the last tragic event to a very tragic week.

The soldiers had proudly stood toe to toe and held their own against a larger, well entrenched, superior force on the now forgotten mountain. The soldiers of 1/22, who took the biggest brunt of the fighting and battle casualties, upheld their motto of "Deeds, Not

Words." Enemy losses would never be known, but with the onslaught of automatic weapon fire, artillery, napalm, and high explosive (HE) bombs, helicopter, and spooky gunships, the numbers had to be well into the several hundred.

Following the soldiers departure from the mountain, B-52 arc light strikes covered the Chu Moor area.

4TH INFANTRY DIVISION
OPERATIONS SUMMARY
21 APR 68—30 APR 68

30 April. At the 1st Battalion, 22d Infantry FSB an outgoing 105mm artillery round struck a radio antenna, resulting in an airburst which killed one US and wounded 12. At AR833230 an MP escort for a Civic Action team was ambushed by an estimated 10 NVA, resulting in two US KIA and one US WIA. At YA996825 Company B, 1st Battalion, 12th Infantry reported one US MIA who became separated from his unit. A search was made with negative findings. At YA881372 a CSF company from DUC CO made contact with an estimated VMC squad, resulting in three CSF WIA.

CHAPTER 12
Memories of a D 1/14 Infantry Soldier
By Jim Brown

LZ Swinger was located to the north of Chu Moor Mountain and the 1/22 firebase, LZ C-Rations, in the Chu Moor Valley Area. LZ Swinger became the base from which 1/14 Infantry conducted operations in the Chu Moor AO.

I WAS IN/ON CHU MOOR. IN OUR UNIT IT'S REFERRED TO AS LZ Swinger. We were inserted under heavy fire and as we were getting off the choppers your people were getting back on. I remember a stack of KIA's and a person trying to keep a poncho over the bodies. Over the years I've communicated with that person, he was one of your medics. I cannot remember his name also I've exchanged e-mails with a chopper pilot that remembers those two days. In an infantry company, the head person is the company commander, the person I was chatting with was a company commanders equal. They lost a chopper and a crew chief lost a foot, so he remembers it well. So much so he was writing a book about the AO.

I was with Delta Co 1/14, our people were told this would be a "Cold LZ", I remember coming in for a landing when I saw black puffs of smoke on the helo pad, the pilot had to abort the landing, and we circled in the air and came back in.

It was like a John Wayne movie, I know you understand. Anyway, after the exchange of troops was made everything slowed down, so much so we were walking around preparing for the night. At some point that afternoon/evening an NVA ran into the perimeter and out again… he was either drugged out or probing, he got away. Sounds crazy but it did happen, bet he's still running with all the lead chasing him (grin).

Later that evening I walked over to another platoon's location to talk to a William Harff. Each day or two he would write down notes for a book he was planning on writing after the war. We chatted a bit then I left to go back to my platoon's location.

On the way there I stopped to talk to our medic and a person named Pete Peterson when mortars came in. One exploded in the air over William Harff causing him to be a KIA; that same round peppered Pete Peterson's back side and knocked down our medic. This was my 1st wound of the war; if Pete hadn't been where he was I would have been KIA. I was hit in the left side with shrapnel and took off running for my bunker, KABOOM, another mortar hit, knocking me to the ground; I was in the open. One of my squad members yelled, "here, get in here." It was a very small bunker with about four people inside. Must have been 25 or so mortar rounds that landed in the perimeter, some were aerial burst; anyway I felt something wet on my back and was coughing. I was wounded again. All the wounds were small but had to be taken out. I remember the gurgling sound in my lung from the second wound. We went into the LZ with approximately 84 men, had two KIA's and 24 or so that weren't wounded or were wounded and could walk.

LZ Swinger was located in a bad spot; the helo pad was in a saddle and marked with sand bags. The next day we were extracted under heavy fire, you could hear the dinks trying to surround us; bamboo was breaking and snapping all around us. I never saw a one; just muzzle flashes that were what I was firing at.

Years later I found out LZ Swinger was abandoned and then re-opened with a battery of 155's placed on the hills on either side of the saddle and the dinks had placed booby traps on the helo pad. It was made out of sand bags, first two birds landed ok but the third one was blown to hell.

I was on the last flight out of LZ Swinger and all I had left was an M-16 with an empty magazine and my steel pot. To this day I don't know how that Huey got off the ground unless it was the hand of God that lifted us out of there. Over the years I have located 65 or so former Delta Co members and put together four reunions; when we talk about it we just say it was a 'bad place.'

Bud Roach—C Company 1/22 Infantry writes:

I was the medic at the LZ. I had rotated out of the field about two weeks earlier after being with C/1/22 for seven months. Charlie Shyab replaced me as senior medic. Then Charlie was wounded (he was the last medic remaining with the company, the rest had already been wounded) and I was sent back in. I carried body bags and my instructions were to get the WIA's and KIA's off the mountain…don't leave anyone. The plan to replace C/1/22 with D/1/14 had already been made.

I rode off the mountain in the chopper with the KIA's. Our medical platoon leader, Bill French, was at the Chu Moor firebase and he said I threw my equipment down and said, "They're all dead." (I didn't remember that). I went in the aid station bunker and stayed a long time. The people in the bags were my friends.

.

CHAPTER 13
Memories of an A 1/12 Infantry Soldier
By Kim Grice

Kim Grice was a member of Company A 1/12 Infantry, when his company was detached and assigned to operational control of 1st Battalion 22nd Infantry during the Battle of Chu Moor. He was wounded during the battle, along with Bernard Mazursky, Thomas McCormick, and Douglas Smith when the four of them were ambushed by the North Vietnamese. Only Grice survived that ambush. The following is his memory of that day.

GOD'S WAYS ARE WONDERFUL!

WESTMINSTER, Colorado—I sincerely believe that Jesus Christ saved my life on May 4, 1968, in the jungle of Vietnam.

There were four soldiers halfway down the side of a mountain that day. We were to give warning to the rest of our unit if the enemy approached. It was about 2:00 p.m. when I heard a noise from some bushes behind us. Sensing that something was not right, I immediately darted for shelter behind a large tree. It was too late. My buddy next to me put his hands up to his face as a bullet come out the back of his head. They had shot all of us in just a few seconds.

I listened for my buddies to shoot back, but there was no return fired. I could only assume that they were mortally wounded. As I lay wounded and disoriented, my total nineteen years of life passed in front of me within seconds. My trained reaction was to get up, locate my rifle, and return weapons fire. But the peacefulness of just lying there for a moment was so comforting.

I knew I had been shot in two places and I began to wonder how serious my wounds were. Silence…and then the enemy was quickly upon us. I dared not twitch as they checked us out, taking whatever equipment they wanted. It was during this time that my prayers for forgiveness were answered and I could feel the Lord put a shield between me and my enemy (only a few feet away). One of my wounded buddies must have moved, so they shot him again. I will always remember being the last person to hear his final desperate cry. As they left, more bullets were fired at us to make sure we were dead. Two of three bullets hit the ground next to me. By this time, I was ready and willing to die.

I was the only survivor of that ambush. The Lord will someday disclose His purpose for me—the reason why He spared me. With anticipation I await the day He calls me home to His kingdom for I truly owe Him my life.

Upon entering the military service, I was given a pocket-sized New Testament by The Gideons. Occasionally, I would read the verses to find hope and strength. While recovering from my wounds in the hospital, my unit commander delivered my possessions to me. To my surprise, the pocket Testament I always carried with me was pierced by a bullet. That Testament helped save my live by deflecting the direction of a bullet that otherwise would have mortally wounded me.

My testimony and that Testament is evidence that the Word of our Lord is powerful. Access to His Word should always be made easy for those who seek it.

Kim R. Grice
Westminster, Colorado
THE GIDEON/April 1989

CHAPTER 14
Remembering Chu Moor

BUD ROACH

For many years I did not know the exact date but I remembered the last week of April 1968. In the past ten years that we have been meeting as a group I know more than I ever did about that week. In the forty years since, a lot of memories have faded but one item was not forgotten. The uncommon valor from many who did not necessarily get the recognition they earned is etched in my memory. An afternoon on Chu Moor Mountain became a defining moment in my life. I am honored to have served with you and my thoughts are with you.

JACK CHAVEZ

That one battle was the defining moment and tragically for many, the last moment in our lives. The last ten years have been a healing time by sharing and remembering together. We are here for one another and give thanks and prayers for those who are not here.

ANDREI PASHIN

I'm sitting safely back at Camp Enari in the Charlie Co. clerk's office when the reports start coming in. I was completely paralyzed, feeling both extremely lucky that Top had pulled me out of the field to take over for the departing clerk and shame and survivor's guilt because I wasn't there. The morning reports during the Chu Moor Mountain days were complex and often initially incorrect. I had to revise them more than once. On April 27, I turned 21 years old, many years older than I had been in 1967 and much, much sadder. My birthdays thereafter were always somewhat marred by the events of April 1968. You will all always occupy a special place in my heart. You were brave young men and I'm proud to have known you.

FRED CHILDS

During the CA onto a very steep hill to establish the firebase, the helicopter tilted drastically and I fell out at a distance of ten feet or so and landed on my head and pack, crushing the steel helmet about 3-4". Like an idiot I got up and said, *"I'm all right"*. Bill Boling remembers my fall and helped me get up. I do not recall any events that took place and later one of the medics told me that it was a good thing I had no memory of that battle. We lost many of our good buddies and many more wounded.

CHAPTER 15
Chu Moor Mountain Battle Deaths
KIA

NAME DOD, DOB, AGE, CITY STATE, BRANCH, RANK, MOS, MARITAL STATUS, ORGANIZATION TYPE, REASON, PROVINCE LOCATION, MEMORIAL LOCATION:

HAMMOND Herbert Lee 4/22/1968 4/11/1948 20 Atlanta GA Army PVT E1 11B1P D/1/22/4th ID Hostile, Died Gun, Small Arms Fire, SVN Kontum 51E 020

BRANCIO David Mike 4/23/1968 9/25/1941 26 Denver CO Army SP4 E4 11B40 D/3/8/4th ID Hostile, Died Gun, Small Arms Fire, SVN Kontum 51E 036

LUCAS Karl 4/25/1968 8/17/1932 35 Redondo Beach CA Army SSGT E6 11B40 D/1/22nd/4th ID Hostile, Died Multiple Fragmentation SVN/Kontum 52E 007

SENA Fred Jr. 4/25/1968 5/19/1945 22 Pueblo CO Army SP4 E4 11B10 D/1/22nd/4th ID Hostile, Died Gun, Small Arms Fire SVN/Kontum 52E 011

CARROLL Thomas J. 4/26/1968 1/14/1944 24 Grand Rapids MI Army CPL E3 11B10 D/1/22nd/4th ID Hostile, Died Multiple Fragmentation SVN/Kontum 52E 015

LAMON Roy Allen 4/26/1968 11/26/1946 21 Liberal KS Army SGT E4 11B20 C/1/22nd/4th ID Hostile, Died Gun, Small Arms Fire SVN/Kontum 52E 020

BUCK Hollis Winfield 4/27/1968 9/17/1931 36 West Paris ME Army SSGT E6 11B40 Married C/1/22nd/4th ID Hostile, Died Multiple Fragmentation SVN/Kontum 52E 028

CAMPBELL James Robert 4/27/1968 11/5/1947 20 Trenton NE Army SP4 E4 11B20 D/1/22nd/4th ID Hostile, Died of Wounds Multiple Fragmentation SVN/Kontum 52E 028

HALL Gary Neal 4/27/1968 11/10/1943 24 Vinita OK Army SGT E4 11D20 A/1/22nd/4th ID Hostile, Died Multiple Fragmentation SVN/Kontum 52E 030

HOMINICK Howard Hugh 4/27/1968 7/7/1948 19 Carle Place NY Army CPL E3 11B10 C/1/22nd/4th ID Hostile, Died Multiple Fragmentation SVN/Kontum 52E 030

JURSZA William Jr. 4/27/1968 2/9/1942 26 Bayonne NJ Army SGT E4 11B20 A/1/22nd/4th ID Hostile, Died Multiple Fragmentation SVN/Kontum 52E 030

SEYKORA, William Joseph 4/27/1968 3/21/1948 20 Owatonna MN Army SGT E4 11B20 A/1/22nd/4th ID Hostile, Died Multiple Fragmentation SVN/Kontum 52E 032

ZEIGLER Roger David 4/27/1968 2/22/1947 21 Yeagertown PA Army CPL E3 11B10 C/1/22nd/4th ID Hostile, Died Multiple Fragmentation SVN/Kontum 52E 032

CASSANO Richard Anthony 4/28/1968 10/7/1947 20 New York NY Army CPL E3 11B10 Married C/1/22nd/4th ID Hostile, Died Artillery, Rocket, Mortar SVN/Kontum 52E 035

HARFF William Henry Jr. 4/28/1968 8/5/1944 23 Kenosha WI Army SP4 E4 11C20 D/1/14th/4th ID Hostile, Died Artillery, Rocket, Mortar SVN/Kontum 52E 037

HOWE Sidney A. 4/28/1968 5/25/1941 26 De Queen AR Army CPL E3 11B10 C/1/22nd/4th ID Hostile, Died Artillery, Rocket, Mortar SVN/Kontum 52E 038

NELSON Lewis Charles 4/28/1968 12/31/1943 24 Seattle WA Army SP4 E4 11C20 D/1/14th/4th ID Hostile, Died Multiple Fragmentation SVN/Kontum 52E 041

NESTER Roger 4/28/1968 10/22/1944 23 Mt. Sterling KY Army CPL E3 11B10 C/1/22nd/4th ID Hostile, Died Artillery, Rocket, Mortar SVN/Kontum 52E 041

ROYALTY Amel Douglas 4/28/1968 9/15/1942 25 Gifford IL Army SP4 E4 11B20 C/1/22nd/4th ID Hostile, Died Artillery, Rocket, Mortar SVN/Kontum 52E 043

YORK Ivol Micheal 4/28/1968 2/3/1948 20 Peru IN Army SGT E4 11B20 A/1/22nd/4th ID Hostile, Died Misadventure/War Accident SVN/Kontum 52E 045

ZIMMERMAN William E. Jr. 4/28/1968 8/5/1942 25 Frederick MD Army 1LT O2 1542 Married C/1/22nd/4th ID Hostile, Died Artillery, Rocket, Mortar SVN/Kontum 52E 045

JOHNSON, Leroy 4/29/1968 1/15/1947 21 Carlisle SC Army CPL E3 11B10 B/1/14th/4th ID Hostile, Died Multiple Fragmentation SVN/Kontum 53E 002

STRICKLER, David Francis 4/29/1968 2/21/1947 21 Falls Church VA Army CPL E3 11B10 B/1/14th/4th ID Hostile, Died Artillery, Rocket, Mortar SVN/Kontum 53E 007

KELLY Seeber J. 4/30/1968 10/14/1948 19 Lake Wales FL Army PFC E3 11B10 B/1/22nd/4th ID Hostile, Died of Wounds Multiple Fragmentation SVN/Kontum 53E 016

WILLIAMS, Robert L 4/30/1968 8/8/1946 21 Washington DC Army PFC E3 11B10 B/1/22nd/4th ID Hostile, Died Misadventure/War Accident SVN/Kontum 53E 026

BONDERER, Thomas Edward 5/1/1968 3/5/1944 24 Chillicothe MO Army SGT E5 11B40 A/1/12th/4th ID Hostile, Died Gun, Small Arms Fire SVN/Kontum 53E 027

BAKER, Reginald 5/2/1968 2/7/1945 23 Chicago IL Army SGT E4 11B20 C/1/22nd/4th ID Hostile, Died of Wounds Gun, Small Arms Fire SVN/Kontum 54E 001

MAZURSKY, Bernard Richard 5/4/1968 1/12/1948 20 Madison WI Army SP4 E4 11B20 A/1/12th/4th ID Hostile, Died Gun, Small Arms Fire SVN/Kontum 54E 037

MCCORMICK, Thomas A. Jr. 5/4/1968 7/28/1949 18 Mattapan MA Army PFC E3 11C10 A/1/12th/4th ID Hostile, Died Multiple Fragmentation SVN/Kontum 54E 038

SMITH, Douglas Bane 5/4/1968 3/18/1947 21 Durham NC Army SP4 E4 11B20 A/1/12th/4th ID Hostile, Died Multiple Fragmentation SVN/Kontum 54E 035

ADAMS, Darrius Wayne 5/5/1968 6/18/1949 18 Ivydale WV Army CPL E3 11B20 A/1/22nd/4th ID Hostile, Died of Wounds Other Explosive Device SVN/Kontum 54E 042

STEADFAST AND LOYAL

POSTSCRIPT
Chu Moor Today

BEGINNING IN 2002, CHU MOOR MOUNTAIN IS NOW PART OF CHU Mom Ray National Park.

Chu Mom Ray National Park has an area of 56.621ha, located in Sa Thay and Ngoc Hoi District, in the west of Kontum province, next to two conservation areas, Virachey National Park of Cambodia and Southeast Ghong conservation area of Laos.

Scientists have noted that Chu Mom Ray National Park has nearly 1,500 species of plants, which has 131 species identified as threatened, such as orchids, and 2,000 species of rare plants like kim giao, bamboo, etc. Scientists have identified 620 species, including 11 species of mammals, 370 birds, 45 reptiles, 20 species of freshwater fish.

Highlights of the National Park boast up to 114 species in the Red Book of Vietnam and the world. In particular, the pasture (Ja Book Valley) on the largest type of Vietnam's areas (over 9,000 ha), in Chu Mom Ray National Park has many hoofed mammals and predators such as tiger, bull, bull elephant, bear, and reptiles.

Visiting Chu Mom Ray, tourists can explore the villages of ethnic minorities such as HLang, Gia Rai, Ko Dong, and Ro Mam. Guests will rest in the hollow or the cool community, enjoy dishes like as rice, Can wine, herbs.

ACKNOWLEDGEMENTS

Bob Simonsen, (deceased) author and former Marine Sergeant (E-5), Vietnam Veteran 1968-69, Company I, 3rd Battalion, 27th Marines, who began this project and created the first drafts.

Jim Demetroulis, D/1/22 Vietnam Veteran who made initial contacts with former soldiers who fought at Chu Moor and initiated this document.

Doug Stanek, D/1/22 Vietnam Veteran who fought in this battle and spent years gathering information, narrations and pictures for his unpublished manuscript, tentatively titled *Seven Zero Bravo*. Doug graciously provided much of the information that he had accumulated and is considered as a co-author.

Rainer Guensch, D/1/22 Vietnam Veteran who fought in this battle, compiled photographs, performed research, and gathered documents.

John McKee, D/1/22 Vietnam Veteran who fought in this battle and provided lists and invaluable information.

Doug Anderson, who produced many of the maps and illustrations.

Michael Belis, C/1/22 Vietnam Veteran, webmaster of the 1-22 infantry website which provided many of the military documents and pictures utilized in this document.

Sam Drake, C/1/22 Vietnam Veteran, contributor and author of Charlie Company location graphics in the Battle of Chu Moor.

Charlie (Doc) Shyab, C/1/22 Vietnam Veteran who was a medic in this battle and with his thoughtful reminders and untiring efforts coordinated the gathering of documents.

John Bobb, C/1/22 Vietnam Veteran who provided insights into a 20 year-old lieutenant platoon leader.

The men of A, B, C, D, Headquarters companies who provided their recollections. Without their assistance this project would never have been completed. My deepest respect is given to the former soldiers who provided their stories and pictures. I know from personal experience that in many circumstances reliving the past can be a very traumatic experience.

In my mind, the real heroes were the Combat Medics. Our infantrymen were trained to drop, find cover, and return fire upon the enemy. The Combat Medics were trained to go to the wounded and provide aid. This was more often than not without the benefit of cover. They are the true heroes.

INDEX